A HISTORY OF
MATHEMATICS EDUCATION
DURING THE
TWENTIETH CENTURY

Angela Lynn Evans Walmsley

University Press of America,® Inc.
Lanham · Boulder · New York · Toronto · Plymouth, UK

Copyright © 2007 by
University Press of America,® Inc.
4501 Forbes Boulevard
Suite 200
Lanham, Maryland 20706
UPA Acquisitions Department (301) 459-3366

Estover Road
Plymouth PL6 7PY
United Kingdom

Library of Congress Control Number: 2007922674
ISBN-13: 978-0-7618-3749-7 (paperback : alk. paper)
ISBN-10: 0-7618-3749-3 (paperback : alk. paper)

For my parents, Ken and Mary Evans

Thank you for instilling the value of education and providing so much support.

Contents

Preface

This book was created because it did not exist—a concise history of mathematics education reform in the twentieth century in summary had never been written. As I work with mathematics pre-service and in-service teachers while researching mathematics education, I have continued to be intrigued with the history of mathematics education; in particular, history that seemed to be repeating itself. Mathematics education tends to swing on a pendulum from "traditional education" including teacher-directed instruction with an emphasis on computation skills to "reform education" including student-directed instruction with an emphasis on problem solving. Understanding and learning from the past is crucial for mathematics teachers and educators as it relates to current reform and teaching.

In this book, I attempt to provide a thorough history of mathematics reform movements as they occurred in the United States during the twentieth century. Each decade of the twentieth century is analyzed within the context of the following conceptual themes: philosophy, mathematics content, teacher education, pedagogy, and assessment, to create a concise history of mathematical reform. Finally, conclusions are drawn as to which reform movements are similar and different throughout the century—depicting which aspects of reform can be seen again. Therefore, all decades are analyzed to see where they were on the pendulum and what aspects may have contributed to the current reform movements led by the *Standards* movement.

It is my hope that this book will help all mathematics educators move forward during these controversial times in education to better serve students. Thank you to the many educators and students of the many traditional and reform movements of the twentieth century. Because of your vision and experimentation, we can continue to learn from you and educate our current and future students better as we strive for perfection in mathematics teaching and learning.

Angela L.E. Walmsley, St. Louis, Missouri, December, 2006

Introduction

Mathematics education has been a popular topic for centuries because mathematics is seen as a cornerstone to education—an essential element of what a child needs to know to work and live in society. However, there has never been a complete history of mathematics education or the reform movements that drove change for children—in particular during its most changing century: the twentieth century. This book attempts to cover those changes by analyzing each decade of the twentieth century, and then connecting historical aspects to current reform. Connections between the decades are made in order to analyze what is the same or different regarding teaching and learning of mathematics.

One definition of reform is that something is wrong or unsatisfactory and needs improvement.[1] Has each reform of the 20th century solved the problems or dissatisfaction of many? It also indicates that whatever the "traditional curriculum" was, needed fixed. Some would argue, yes, that the traditional curriculum does need some help as it often deals with "how" to do mathematics but not "why" something in mathematics is done. But has the public generally been happy with either the traditional curriculum or the reform movements toward change?

What is "traditional" and what is reform? Typical practices of "traditional" mathematics teaching include the use of mass produced textbooks by commercial publishers; instruction is mainly teacher driven; teachers have a great deal of autonomy in how and what they teach; teachers who teach mathematics do not necessarily feel they are strong in the subject; and math curriculum tend to emphasize computational skills at the expense of problem-solving skills.[2] These "typical" characteristics describe teachers who may use "direct instruction" as their only means of instruction, and stress the teaching of basic skills. This teacher-led instruction has, through the years, been criticized for its "boring" and overly structured method. On the other hand, "reform" movements have varied but often stem around the focus of problem solving at the expense of basic skills using a teaching philosophy of constructivism. There are various forms of constructivism, and what many people think of when they hear this word is students constructing their own knowledge in a student-directed classroom. Constructivism has been criticized for being too ambiguous with little direction in the classroom. These explanations can vaguely be the key components of the dichotomy described next.

During the twentieth century, mathematics reform has swung on a pendulum between "drill and practice" and "discovery learning."[3] The NACOME report alluded to the problems of dichotomies within the mathematical community. People believe that they must choose between

> the old and the new in mathematics
> skills and concepts
> the concrete and the abstract
> intuition and formalism
> structure and problem-solving
> induction and deduction.[4, p.136]

Why do educators need to choose one end of the pendulum? Surely a "middle-ground" would make the most sense for most children. Dewey, in 1902, noticed these two ends of the spectrum with one group wanting traditional education taught to students step by step and another group wanting a more student-centered approach.[5]

Another problem with swinging back and forth is that people fail to actually study the history of the reforms and learn from what has come before. "If we want to understand our profession today, if we want to guess what will happen tomorrow, we ought to study yesterday."[6, p.663] In 1978, James Fey stated,

> It is not uncommon to hear mathematicians criticizing the schools in much the same way that their predecessors complained 20 years earlier; recommendations on the proper direction of educational practice and policy seem to reflect no awareness of, or curiosity about, the historical experience with similar problems or conjectured solutions.[7, p.346]

Therefore, by analyzing each decade, educators can take from it the best parts of mathematics reform and relate it to education today. Therefore, in the next century, rather than continually repeating patterns, we can learn from history and continue to develop a strong mathematics education agenda.

In developing each decade of the 20th century, the author will first provide a short description of any education trends for that decade. This will be followed by a concise history of what was occurring nationally and politically as these issues sometimes indirectly and directly affect what is done in public education. Also, the author will present each decade of analysis in the mathematics education contextual framework of philosophy of education in general, mathematics content, mathematics teacher education, pedagogy, and assessments.

1900—1910

The turn of the century saw a growth in the number of children who attended school for longer periods of time. Only one in ten attended high school in 1890, but by 1920 one in three was attending high school, and by 1935, two in three were attending high school.[8] The century also started with the establishment of the 1893 "Committee of Ten" report which proposed what eventually became the standard high school curriculum.[9]

Historical context

From the beginning of the 20th century until World War I began the following decade, the country moved towards change in the "progressive era."[10] In the early 1900s, many children were part of the labor force or agricultural force rather than spending their days in schools. Child labor laws came into play in this decade, but a majority of children were forced to work rather than attend school. In fact, a majority of children attended school only through the eighth grade with few attending high school and even fewer attending college.[11] It was during this time as well that women were fighting for the right to vote. While that right was not granted until 1920, it would appear that this could inhibit girls' education in the early twentieth century. The role of the progressives was to work towards greater democracy and social justice for all. At the beginning of this movement, many cities were filled with slums, and almost half of all industrial workers lived in poverty.[12] It is during this decade that some of the following initiatives began to develop: prohibition of the sale of alcohol, restricted child labor laws and work hours, and compensation for injured workers.

Philosophy

In 1900, the majority of students enrolled in high schools were studying traditional academic subjects such as Latin, sciences, mathematics, and history.

At the turn of the century, the popular method for schools was to use the "drill-and-recitation" method. The 1900s saw the start of the progressive movement in education—often associated with John Dewey. Dewey had recommended that schools be used as instruments of social change.[13] Dewey had opened his laboratory school shortly before the start of this decade, and he espoused that the school could serve the functions of a community to completely educate a child. He also felt that students needed to learn about things they were interested in and could relate to, and take the lead in their own education based on their experiences and imaginations. In turn, Dewey believed that children could learn democracy through their own interests and grow into citizens who could develop the United States democracy. In practice, progressive education, as a theory described above, worked well in elite schools that consisted of middle class to higher class students, and did little for immigrant children.[14]

This decade saw changes in education as Americans felt the traditional college preparatory schools were not adequate in serving the needs of all children. Firstly, a number of families were moving into cities, and this increase in urbanization gave rise to developing schools to educate many of these students. However, controversy rose between whether students should be placed on a "vocational" track or an "academic" track. While both served to replace the "classical" curriculum with a more modern one—they were definitely different. Nevertheless, progressive education was modeled on the following four points: 1) the traditional curriculum should be replaced by one based on the interests and needs of the students; 2) learning should be activity based rather than rote; 3) schools should reflect their communities and social conditions; and 4) a primary focus of school should be in resolving social problems.[15]

As stated earlier, this decade saw a large increase in the number of children attending school. At the end of this decade, approximately 82% of secondary age pupils were enrolled in school compared to 74% at the beginning of the decade and 57% in 1890.

Mathematics content

The Committee of Ten, who recommended an outline of school mathematics in the late 1800s, had established the pattern of algebra, geometry, advanced algebra, and trigonometry.[16] The Committee of Fifteen on Elementary Education also called for the use of arithmetic through grade six, with algebra being the emphasis in grade seven and eight.[17] At this same time, the Committee on College-Entrance Requirements made recommendations along the lines of the Committee of Ten: algebra and geometry should be integrated into education at grade seven progressing through grade twelve with advanced algebra and trigonometry in grades eleven and twelve. This pattern has emerged as the most prominent pattern in the United States making it the only country to not have a unified mathematics curriculum throughout school.[18] While many argued for unified mathematics even at the turn of the century, it has never been totally successful in the United States.

However, at the time of the recommendations, the majority of those attending high schools were college bound. Most educators at the time felt that the study of mathematics was for two purposes: classical humanism and mental discipline.[19] Classical humanism meant that the study of mathematics has historically and traditionally been part of future preparation. Mental discipline meant that mathematics study was necessary to improve and exercise the mind. Therefore, much of the curriculum was based in arithmetic, and those who entered high school normally took the following progression of courses: arithmetic, algebra, geometry, and trigonometry. Sometimes trigonometry, pre-calculus, and analysis were taught at the college level only.[20] The courses, while not being integrated, were also not tied to other subjects.

Teacher education

At the beginning of the 20th century, some teachers began teaching with only a high school diploma and no other training.[21] While other teachers were being taught in two year normal schools, the trend was gradually changing to four years of education for the training of teachers. With this change, a national effort began for normal schools to change into teachers colleges. It is at this time that the three forces within a program were established: the study of academic subject matter; the study of the profession of education; and practice teaching in a school.[22] However, there were critics stating that teachers did not have enough mathematics in their backgrounds. During this decade, a mathematics teacher probably graduated from college with only about one year's work further advanced in mathematics than the level he or she was about to teach.[23]

Pedagogy

The role of progressive education was partly to engage students in their own learning. Hence, during this and the next decade, schools tried to make the classroom more "student friendly" by having students work together in groups with hands on activities. Before this time, children were required to memorize lessons, recite them flawlessly, all while every other child sat still and quiet in small and crowded classrooms.[24] Mental mathematics was an important component of mathematics education at the turn of the century. The pedagogy of progressivism was to allow children the freedom to move their bodies and minds as necessary to advance learning for the whole child. Many would argue that progressive education was the first time teachers were asked to really analyze and understand different types of pedagogy. Some would argue that while progressive pedagogy worked well with middle and upper class children, it was difficult for poorer children because they lacked a variety of life experiences and needed more structure from the teacher on a daily basis.

Assessments

Assessments at the turn of the century depicted the standard rote method of learning by asking students to complete standard, quick, mathematics computational problems in basic addition, subtraction, multiplication, and division. Although not regularly performed in many schools by any means, it was at this time that some of the first large scale studies involving standardized testing began.[25] Due to the popularity of the invention of psychological and intelligence testing during the early 1900s, many students saw the beginning of achievement tests and normal curve placement.

1910—1920

From the beginning of the century, Dewey urged a change in public education from educating the elite to educating all in a child-centered fashion.[26] World War I and the publication of the Cardinal Principles of Secondary Education enabled a progressive climate during the end of this decade into the next two.[27] These principles stated that there were seven basic goals for education: health, fundamental processes, home membership, civic education, vocation, leisure, and ethics. The establishment of these areas in schools would support democratic citizens who could leave productive lives. Nonetheless, this broader curriculum did not occur in many schools and the traditional curriculum remained the norm.[28]

Historical context

As the progressive movement continued into this decade, President Wilson concentrated on regulating the banking business to help ensure a lower cost of living for citizens. Wilson also helped pass labor laws against child workers.[29] While congress passed many acts to help the United States move forward during this decade, the decade faced many hardships as World War I developed beginning in 1914. In 1917, The United States entered the war, which ended by 1918. It was at this time that many individuals paid higher taxes and bought war bonds to fund the war.[30] America experienced anti-German feelings which spread to German-Americans. Many Americans were arrested for speaking out against the war which was seen as unpatriotic. After the end of the war, while Wilson continued to try and involve the US in world government through the League of Nations, Americans were more interested in getting back to their lives at home. In fact, they became leery of immigrants and supported deportation of a number of innocent foreign-born political radicals. This also led to restrictions passed for immigration—severely limiting numbers of immigrants and where they came from.

Philosophy

As the progressive education movement continued during this decade, educators saw two forms. First, progressive education was seen in many schools educating children from lower-class families, and much of the emphasis was on vocational and technical courses.[31] This reality was caused by business leaders who wanted students prepared to work in the industrial age, and educators who wanted the curriculum to align with the needs of society in this industrial age. This is also the decade that saw the expansion of junior high schools—mainly to shorten elementary programs and get students in a correct "track" during the junior high school years—leaving only those college bound for academic high school preparation.

Another form of progressive schools was seen as experiments aimed at educating students from the upper classes. Dewey supported students' understanding through their own experiences, which would in turn lead to new experiences in a continuous and progressive cycle.[32] It was also after the war that the emphasis in education switched from the ideals of John Dewey (who wanted to use schools to create social change) to helping children develop emotionally.[33] "Child Centered" became a theme at the expense of learning traditional academics. The outcomes many agreed were children with a lower work ethic who didn't have enough academics. This was seen as a problem because by the end of this decade, approximately 88% of high school aged children were enrolled in secondary schools.[34]

Mathematics content

During this time, there was a movement to change mathematics learning to include making connections between different types of mathematics and unifying them together.[35] This movement was led by Moore, a mathematician from Chicago, who called for unification and ties to the sciences where students could participate in lab work seeing mathematics in action. While unification became popular in junior high schools, it never really took hold in high schools.[36] However, at the same time, a "social efficiency movement" was taking shape that emphasized mathematics as a drill-and-practice subject; therefore, many schools moved directly from the traditional mental exercises of mathematics to this concrete drill-and-practice method based on practical value.[37]

The Cardinal Principles was a report in 1918 that basically stated that not all students should be required to take the "traditional" mathematics courses of algebra and geometry. Because more students were going to high school, and many were training in vocational tracks, the recommendations were made that those students could take just one year of mathematics—even a generic mathematics course—to graduate from high school.[38] While the *Cardinal Principles* took a position for education for all, the focus on higher level mathematics in schools declined.   Those college bound still took algebra, geometry, and trigonometry, but those not college bound usually found themselves floundering in these courses if they elected to enroll.   Therefore, despite any initial progressive ideals, what eventually became associated with the progressive movement was a shift in mathematics content to only subject matter that the average citizen may need in an industrial job.

Teacher education

It was during the progressive era where teacher education changes were made.   The attainment of a four year baccalaureate degree was becoming standard—this caused normal schools to either close or develop four year programs themselves.   A common complaint of the normal schools was that they stressed teaching with less emphasis on subject matter.[39] Other students attended four year universities to complete their teaching degree.   In these institutions, a subject matter was prescribed with a few teaching courses added.   Therefore, the degree stressed both pedagogy and academics.   The National Council of Teachers of Mathematics also started in 1920 consolidating resources into one voice for school mathematics.[40] It was also at this time that summer courses and schedules were designed for in-service teachers—mainly through normal schools. These changes helped increase the professionalism of teaching.

However, in 1918, a report stated that the United States could not offer high levels of mathematics in schools because it lacked teachers highly trained in mathematics as well as individuals with strong mathematical backgrounds who wanted to become teachers. Furthermore, there was a constant complaint that with the increase in secondary schools, the focus of education of teachers was on secondary teachers and not elementary teachers.   In fact, during this decade, while most of the secondary teachers had four year degrees, some of the elementary teachers still only had two year degrees.   It is during this time that the mathematical competence of the elementary teacher was first questioned.

Pedagogy

As teachers changed their teaching methods from rote learning and memorization to a more child-centered curriculum, students began to focus more on social aspects of interacting well with each other and with the teacher.   The Progressive Education Association (PEA) organized in 1919, and their major principles included (among others): "Freedom to develop naturally; interest the motive of all work; the teacher as a guide, not a taskmaster; and the progressive school as leader in educational movements."[41, p.119]

Thus, pedagogy and the curriculum became dominated by social interactions among students.   Dewey espoused learning mathematics with activity.[42] However, in reality, the majority of schools during the progressive movement did not change totally, and a teacher-dominated classroom and curriculum remained popular.[43] Furthermore, drill continued to be the main means of instruction.[44]

Assessments

A common complaint about assessments in general despite any decade it occurs in, is that students should be fairly assessed based on the content of material they were taught. Furthermore, it is difficult to compare various students who might have been involved in trial programs or in traditional schools because assessment cannot possibly be the same for both groups.[45] Therefore, while progressivism was popular at this time, assessment of students who participated in a progressive program against those who did not cannot adequately assess the work of both groups of students fairly.

Psychological testing that was becoming popular at the turn of the century, was well established by 1920 and was being used regularly in education.[46] It was during this decade that school saw a huge increase in enrollment (especially since the turn of the century).[47] Therefore, psychological tests allowed many suddenly new students to be placed accordingly to their interests and abilities.

1920—1930

Historical context

The first major accomplishment of 1920 was the final ratification allowing women to vote. The presidency changed hands to the Republicans who vowed a return to "normalcy" for the United States.[48] Immediately post-war, the United States economy was fairly strong. However, by the mid 1920s the economy began to boom and the country enjoyed a strong economy until the end of this decade. Due in large part to the growing productivity of many businesses, Americans enjoyed a better lifestyle during this time than ever before.

Many Americans bought cars, radios, refrigerators, and vacuum cleaners for the first time during this decade.[49] They also began enjoying life by attending movie theaters, parties, dances, etc.[50] This was also the time of prohibition—leading to many speakeasies and other places of entertainment. This decade was called the "Roaring Twenties" with individuals leading a more open or vibrant life. Between 1921 and 1929 an unusual degree of economic stability and an exceptionally low unemployment rate accompanied high levels of income.[51] Citizens enjoyed the move theater as well as jazz clubs—women cut their hair short, wore shorter skirts and silk stockings. Some were afraid that "morality" had left America. It is during this time that factories and companies began offering pensions, installing safety measures, and promoting a positive working environment. The decade ended with a "bang" when, in October 1929, the stock market crashed leading to the Great Depression in the 1930s.

Philosophy

At the beginning of this decade, students had shifted from the majority taking "traditional academic subjects" to many taking subjects of a vocational matter such as typewriting, home economics, agriculture, etc.[52] Music, art, and physical education were becoming more popular also. It was during this decade that America saw a shift in education. First, the per capita amount spent on children in schools jumped from $24 in 1910 to $90 in 1930.[53] Free elementary education was now established throughout the country, many more students began to attend high schools and colleges, and illiteracy dropped from 7.7% to 4.3%.

The progressive education movement had impacted many Americans in the following way: creating two tracks of education, one vocational and one academic, and many children were placed on the appropriate track after taking the infamous IQ test that was further developed by Louis Terman.[54] Elementary schools provided a similar education for all students—the changes seemed to focus more in the secondary school. However, by the end of this decade, many more Americans were being educated compared to the beginning of the progressive education movement around the turn of the century. Most attended through the eighth grade with many continuing into high schools (although the drop out rate was high here as students left for the work force). What was essentially happening; however, was the large numbers of students in high schools caused a push into the white-collar work force more than it had before.[55]

Mathematics content

Besides the *Cardinal Principles* mentioned previously, another report, the *1923 Report* stipulated that algebra and mathematics was important for all, and that the focus of school mathematics needed to be higher standards in mathematics.[56] This report solidified the junior high and senior high mathematics curricula. This report remained the major influence in mathematics curriculum until the CEEB (College Entrance Examination Board) report in 1959. In fact, *The Reorganization of Mathematics in Secondary Education* stressed that a general mathematics approach should be available to grades 7-9 that included statistics, trigonometry, algebra, and geometry.[57] While this report had some influence, the *Cardinal Principles* had already become too ingrained in society to alter mathematics education.

As the progressive movement continued, it became apparent that two schools of thought about mathematics continued to evolve: one for the college-bound to take a more traditional set of courses, and one for everyone else that focused on only what mathematics they would need for their future. This "social efficiency movement" caused mathematics focus to decrease in elementary and junior high schools as the only mathematics taught was that which was seen as practical. In the high schools, mathematics requirements were removed; this is partly due to the fact that so many "ordinary" children were now attending secondary schools and needed to be trained for life and jobs rather than the traditional academy.[58] Therefore, the curriculum focused on arithmetic with little understanding of why algorithms worked or where they came from.

Teacher education

It was during this time that teachers became more involved with university leaders and others to analyze and develop curriculum.[59] The National Council of Teachers of Mathematics was founded in 1920.[60] The Mathematical Association of America published a report in 1923 asking that secondary mathematics be taught for understanding with less focus on drill-and-practice.[61] In the early part of the century, there were three patterns of what courses future teachers should take: those in pure mathematics, those in applied mathematics, and those in pedagogy or the teaching of mathematics.[62]

The traditional normal schools focused mainly on pedagogy leaving many teachers with a little background in mathematics. It was at this time that there was a growth of teacher colleges that had evolved from normal schools in the United States.[63] In fact, the most common complaint of elementary teachers has been that they did not perceive themselves in having a strong enough background or understanding in mathematics content or instruction.[64] Teachers at this time were taught to be generalists and not have a specific area of content expertise. Furthermore, teachers were inclined to teach mathematics the way they were taught because they were not strong enough in mathematics content or instruction to alter methods.

At the beginning of the century, only one state had exercised control for certifying teachers. By the end of the 1920s, at least ten states were certifying teachers and offering life teaching certificates to teachers who attended in-service education.[65]

Pedagogy

By the mid-1920s, progressive education was considered the wave of the future—especially with its child-centered view.[66] Many Americans didn't realize that it had been in effect at many private schools for a number of years. Some of these techniques were individualized instruction, group projects, projects based on student interest, and creative activities to reach academic goals of learning.

Because of the increase in the numbers of students attending school for general education (as opposed to the smaller numbers of preparatory students in the past), many more teachers had to be trained; furthermore, training methods had to be reexamined because of the change in the average student.[67]

Assessments

By the end of this decade, testing was used regularly to determine future paths for students. Most educators began relying on these tests for evaluation of students in general knowledge. Therefore, the tests were not used to assess teaching, but instead were used to assess general intelligence that was not necessarily affected by school. It is estimated that as this decade ended, there were approximately 1300 achievement tests and 400 intelligence tests available for schools to use.[68] Because multiple choice response became popular at this time from its ease of scoring and objectivity, students were asked to perform many rote computations, and basic skills testing remained popular.

The first SAT test was given in 1926 with 8,040 students taking the test. This examination became the standard by which high school graduates were measured upon entrance to university or college. This number can be compared to 1986 when over 1 million students took this exam. This decade set the stage for the examination and steady increase in the numbers of students to attend college.

1930—1940

By 1932, eight million Americans were out of work, and some educators asked progressive educators to work hard at helping students become economically successfully and bring about change.[69] Because of the lack of employment, many students stayed in school impacting secondary school mathematics. But, at the same time, fewer students enrolled at universities. Therefore, progressivism continued to be popular during this decade as many saw this form of education as a "cure" for the economic problems in the country.[70] It was a philosophy that citizens felt could eventually provide students with abilities to function in a democratic society. In reality, this "Social Reconstructionism" never became a major focus.[71]

Historical context

As the 1930s began, a number of Americans were unemployed, and many others never recovered the money lost in the great stock market crash. Individuals were amazed at this time when, having always believed that hard work would earn a decent living, they were not able to find work or provide for their families. When Franklin D. Roosevelt took office in 1933, he began programs that became known as the "New Deal." The New Deal introduced many social and economic reforms to try and stabilize the poor economy at the time. Roosevelt's progress included reforms in banking, work relief for unemployment, government national building projects, relief and economic stability for farmers, and labor organization.[72] It was during the mid-1930s that the government concentrated on creating Social Security to benefit older, disabled, unemployed, or poor Americans. It was also at this time that conflict began to develop in Europe, and Hitler and the German Reich began to rise in power.

This was a time for promotion of the arts with the Federal Art Project, the Federal Music Project, and the Federal Theatre Project.[73] Americans still enjoyed going to the movies and listening to the radio as pastimes. It is in the late 1930s that more child labor laws were enacted—prohibiting children under the age of 16 to work in any industry other than farming. Prohibition was seen as a difficult law to enforce, and it was appealed by a new Constitutional Amendment in 1933.

Philosophy

By the end of this decade, approximately 93% of all high school aged children were enrolled in high schools. In fact, many would argue that public high schools did not become a mass institution until this decade.[74] James B. Conant, a spokesman for education during this and the next couple of decades, was president of Harvard University from 1933 to 1953. It was in this role that Conant promoted his ideas of what public education should encompass. Conant believed in tracking or streaming—that is placing students in educational roles that would suit their intellect the best, despite their socio-economic background. Conant, however, was in favor of using testing as a means to determine who was the strongest academically. He continued to support that only the best of the best should enter higher education.

During the depression era of the 1930s, the government became more involved in education by constructing more schools, providing free lunches for poor children, and providing part time work programs for high school and college students—all stemming from Roosevelt's New Deal initiatives to help Americans out of the poor economy.[75]

Mathematics content

The Great Depression left many students without the option of work; thus, more stayed in school for a longer time. Because of this, junior high schools and high schools often developed new mathematics courses similar to "consumer mathematics," where students learned life mathematics skills such as balancing family assets, renting a home, etc.[76] The Progressive Education Association reported that educators should provide a general mathematics curriculum to meet the needs of students to live in a democratic society. In fact, the "core curriculum" of the time wasn't based in academics at all—it focused on students' experiences and problems they might encounter.[77] With the effects of the depression, focus on education for all students further caused an arithmetic based mathematics for all ideal with little focus on higher level mathematics.[78] What continued to be strange, however, was that many students continued to enroll in traditional academic courses (when allowed) despite their vocational "track" and the fact that they would not attend college.[79]

Many mathematics educators were not in favor of the majority of students taking "life skills mathematics" courses, but educators as a whole disregarded changes to this now status quo. There was a movement to postpone basic arithmetic teaching and other skills until children were more advanced.[80] This practice, in general, continued into the 1950s. Much of the research of the times indicated that students could not necessarily transfer learning easily; therefore, the option to teach practical and basic skills continued to be the favored options for the mathematics curriculum.[81] In reality, students were taught arithmetic without understanding and seeing connections to other aspects of mathematics or other subjects. Nevertheless, the mathematical community continued to call for teaching for understanding. In particular, they felt that because of the introduction of junior high schools, that geometry and algebra should be introduced in seventh and eighth grades leaving more room in high school to expand the mathematics curriculum.[82] Those on a college bound track continued to take geometry, advanced algebra, and trigonometry in high school.

Teacher education

The 1930s was an era of "aggressive professionalism" in education with standards for certification and teacher education rising.[83] In fact, some progressive educators felt that normal schools were doing an inadequate job of preparing teachers—thus teacher education became a focus for the first time. Curriculum reform took place during this decade also due to the number of teachers and administrators who began to attend conferences. Thus, teacher education was framing into more of a profession than it had before.

Regarding mathematics specifically, teachers of seventh and eighth grade needed more training in mathematics with the development of the junior high schools and the pushing down of algebra and geometry into these schools.[84] Also, a report in 1935 stated that high school teachers should take college algebra, analytic geometry, calculus, and a critical examination of Euclidean geometry, history of mathematics, equation theory, some modern algebra, advanced calculus, and some work in a science.[85] In addition, a course in teaching mathematics was also recommended. Therefore, universities began offering courses specifically for these teachers.

Pedagogy

Teacher pedagogy in the 1930s continued to focus on education for the child, and was based on the theory that schools had a responsibility for changing society.[86] The focus was on teaching children rather than teaching subject matter or academics. This came to be known as "the activity movement," but followed along the same ideals of progressive education such as project based learning. The 1930s saw the beginning of the movement for teaching for learning or understanding. It was the first time during this century that teachers were asked to be creative and non-repetitive and avoid the automatic use of rules.[87] Nevertheless, because of economic depression, education became more focused on training children specifically rather than letting them choose their own path of learning.[88] There were also no funds to try new pedagogy or new techniques; thus teachers remained more stagnant during this time.[89] Thus, students were being taught basic mathematics.

Assessments

In the 1930s, The Progressive Education Association initiated a study known as the Eight Year Study that evaluated students who had graduated from both progressive and traditional schools in order to compare their achievements.[90] The results showed that students who had graduated from the progressive schools earned a slightly higher grade point average and slightly more academic honors, were thought to have better problem solving skills, and tended to specialize in the same fields as the traditional students. While some assessments did center on problem-based learning, a majority continued to be based in basic skills and quick, solvable problems.

1940—1950

The Joint Commission report of 1940 tried to increase the public's value of mathematics; however with World War II, the need for the public became enough knowledge for mathematically strong soldiers. The post war years brought more emphasis on science and mathematics as soldiers, politicians, and the public saw that the United States was not as technologically advanced as it should have been during World War II.[91] However, much of the decade was spent educating students for "functional competence" rather than increasing sharply the level of mathematics offered.[92] Nevertheless, researchers at universities were beginning to voice their concerns about the lack of high levels of mathematics taught in schools. Thus, this is the decade that framed the initiation of the "new math" movement.

Historical context

This decade began with tension as World War II developed in Europe. While most Americans felt neutrality was key, their opinions changed with the bombing of Pearl Harbor in 1941. The United States was preoccupied with the war in every way until surrender in 1944 in Europe and in 1945 in Japan. Many Americans who entered the military as men between eighteen and forty-five were available for war because of the draft.[93] Women also became involved in the working force because of the number of men overseas. Also, America was occupied with making ammunition and products directly for the war cause. Much of the war was paid for by increasing taxes on citizens. Goods were rationed—including sugar, coffee, gasoline, tires, etc. President Truman ordered the use of atomic bombs against the Japanese—acts that are considered the reason for the end of the war. Congress also promptly approved the organization of the United Nations after the end of the war. However, throughout the 1940s, the struggle with Russia and the Cold War continued.

During and after the war, many Americans began to move west of the Mississippi river for higher wages with more opportunity for jobs. After soldiers returned from the war, America saw the beginning of the "baby boom" era. In the 1940s, America grew by 19 million people compared to 9 million in growth during the 1930s. The GI Bill provided money for veterans for education, housing, medical treatment, and other loans. Overall, the postwar era saw Americans enjoying life again as they socialized, traveled on aircraft, and bought modern appliances for their new homes. Americans were moving into a lifestyle that is more characteristic of modern day.

Philosophy

The GI Bill (GI meaning "government issue") allowed the opportunity for many soldiers to attend college—something that they would not necessarily have been able to do because of the poorer background from which many came. The government initiated this bill to try to avoid high levels of post-war unemployment, but veterans showed their abilities to perform well in the liberal arts.[94] By 1947, almost half of all college students were World War II veterans.[95] Also, following World War II, critics stated that Americans were not academically challenged and not learning math and science to their potential. One critic, James B. Conant, stated that America was ignoring its brightest children, and at the same time, not promoting English grammar or foreign languages enough.[96]

Instead, it appeared that schools were focused on educating students to become citizens—stressing social skills at the expense of academics.[97] In fact, in 1944, the National Education Association published *Education for All American Youth,* which basically stated that schools would assess students into academic or vocational tracks, and then children would proceed as necessary by these recommendations.[98] Schools would educate and prepare students for their selected "social order." These students would study a "common learnings" core curriculum. By the end of the decade, many critics stated that schools were lowering standards for their children. In particular, military recruits had to be re-trained in arithmetic and other mathematics in order to perform necessities for the war.[99] Many saw that in order to be a successful country during wartime or peacetime, more mathematics, and not just "practical mathematics" needed to be taught to the average student.

Mathematics content

In 1940, a joint commission of the Mathematics Association of America (MAA) and the National Council of Teachers of Mathematics (NCTM) published a yearbook focusing on secondary mathematics education. In it, they called for students ready to take calculus when they entered college—this would require students to have a background in algebra, geometry, and trigonometry as well as some kind of twelfth grade mathematics course.[100] Furthermore, while this commission focused on a college bound track, it also stated that a track for non-college bound should include more than elementary level arithmetic.[101] The commission stated that the track should include a spiral curriculum in which students learned more mathematics each year related to consumer mathematics.

Admiral Nimitz declared that a large majority of naval recruits could not complete simple mathematics assessments because of lack of sufficient preparation in school. For a short time, emphasis in schools was in mathematics and the importance of "traditional" higher level mathematics such as algebra and geometry for more students.   Some schools introduced new courses that combined elements of mathematics for students who would soon graduate and possibly join the armed forces. Schools began to use military mathematical examples to help with recruitment, and many mathematics educators promoted the focus of understanding of mathematics as necessary for engineering and military causes.[102] The importance of mathematics was made the focus for many average Americans, and many saw it as patriotic to promote mathematics since in the end, it would support the war effort on all accounts.

The mathematics content was divided after World War II between those who felt that advanced academic mathematics should have been taught and is necessary for advancement in science and the military efforts, and those who felt that "consumer mathematics" should still be the focus after the war since students only needed to understand enough mathematics for vocational jobs and living a "regular" life.[103] The National Council of Teachers of Mathematics published an account on Post-War Plans in 1944.   The commission who published the plans stated that three tracks should be developed: the same traditional academic track for high achievers in mathematics; a middle track for those students who will use mathematics in their careers; and a social track for students who may not use mathematics in a wide context but should have a background in every day mathematics.[104] This commission published three reports, and set the tone for the rest of the decade.

While schools saw a brief change in focus on what mathematics should be, in reality, little changed in the schools. By the time the war ended, schools began to focus again on teaching only what students would need for mainstream life in America.   The standard curriculum of teaching practical mathematics skills for life became known as "life adjustment education," and became the quasi-official education policy for the US.

Teacher education

By the beginning of World War II, teacher education was primarily a four year degree obtained at a state institution with state regulations.[105] However, critics claimed that, while teachers were instructed in pedagogy or teaching methods, they did not receive an adequate education in subject matter. In fact, that was the single biggest complaint about elementary teacher preparation programs.[106] The joint commission of the MAA and NCTM also stated that high school mathematics teachers were not adequately prepared to teach mathematics because the level of courses required for a license was too low.[107] In particular, some states didn't require any college level mathematics, and others only required a one year course.

The MAA and NCTM called for all high school teachers to have taken mathematics to the level equivalent to calculus by the time they finished their college education, and every secondary mathematics education department should have one teacher highly trained in mathematics. In particular, they stated that all teachers who would be teaching mathematics at the secondary level take, in addition to education methods courses, college algebra, analytic geometry, six semester hours of calculus, geometry, advanced algebra (including finance and statistics), the history of mathematics, and a course in a related field such as physics. Those teachers who specialized in mathematics should also take advanced calculus, differential equations, and additional work in geometry, algebra, and a related field.

However, by the end of this decade, very few states were responsible for teacher certification of high levels of mathematics. Few states required any training in mathematics methods or content for elementary teachers, and only two-thirds of states required some type of training in mathematics for initial certification and most did not require a methods course.[108]

Pedagogy

Public schools had become agencies dedicated to socializing students, teaching them proper behaviors, and encouraging conformity to the norms of social life.[109] Therefore, teachers felt they were succeeding in preparing students to life and adding to the democratic character of the United States. However, at the same time, parent and community leaders were unhappy with the academic preparation of students graduating from high schools. In particular, many Americans could see, after World War II, that the need for strong academics with entry into a technological and "white-collar" world would be important. These same people were skeptical that the current public school was doing enough to prepare students for a more advanced world. Therefore, society saw a decline in the progressive education movement during the war years, with official disbandment of The Progressive Education Association in 1955.[110] Thus, the next decade saw a shift away from this progressive pedagogy to more academic centered.

Assessments

A complaint of the average college freshmen in the 1940s was that he or she could only do geometrical proofs and had inadequate skills to solve application problems.[111] This could be explained by the fact that during this period, multiple choice and assessment tests were popular because of their ease in giving and grading. Thus, open response or open ended questions were not seen as favorable as these "new ways" to grade. By the mid-1940s, alternative, mechanical means of scoring were being investigated which prompted more use of these types of tests.[112] Because of the popularity of these tests, standardized tests were refined and publicly accepted as the major means to continually test students into the 1950s and 1960s.

1950—1960

The beginning of this decade saw a continued decrease in the number of students taking college-track courses such as geometry and trigonometry. In fact, even as late as 1954, only 23% of high schools offered solid geometry as a course.[113] However, the "new math" movement started during this era; but really became favorable with the launching of *Sputnik* in 1957. (For a complete history of the new mathematics movement, see *A History of the "New Mathematics" Movement and its Relationship with Current Mathematical Reform* by Angela L.E. Walmsley, published by University Press of America, 2003). It was at this time that more federal funding became available for mathematics programs, and many institutions and universities started their own "new math" programs. In fact, the "new math" movement really began with the University of Illinois project in 1951, and continued until federal funding ceased most programs in the 1970s.[114] The most popular "new math" project that became known was the School Mathematics Study Group because it received the most federal funding and produced textbooks quickly.[115] The "new math" movement was an exciting time in mathematics education because the focus of mathematics was in the public eye for at least ten years.[116]

Historical context

The Cold War continued in earnest throughout the 1950s while Senator McCarthy led his campaign to rid the United States of Communist supporters and Americans fought for three years in the Korean War. Another unique aspect of this decade was the increase in attention to space programs. It appeared that American and Soviet citizens felt a competition between the two countries to advance technology into space travel. It was *Sputnik*, the Russian satellite launched in 1957 which led to a spark in mathematics education among average Americans.

Nevertheless, America's economy began to boom. This can be attributed to a number of things, but mainly that there was a rise in defense spending as the Cold War escalated, and many Americans began to enjoy more working and middle class life by purchasing homes, campers, boats, television sets, and automobiles.[117] Advertisements and TV shows became popular during this decade.[118] Purchasing was not a problem since the idea of "credit" was becoming widespread in America. With businesses expanding, America saw fewer "blue-collar" jobs and more "white-collar" jobs offering benefits and a guaranteed annual wage. As a result, city life began to dwindle, and suburbs and shopping centers became very popular. This was all made easier as highways developed and the interstate system formalized. High schools flourished creating their own sub cultures of youth—many who were listening to rock and roll, smoking, drinking, and driving their own cars.[119]

The babies from the "baby boom" era became teen-agers at the end of the 1950s—this led to two counteracting groups. One group was teen-agers with money to spend weekly on merchandise, parties, etc. The other group was teen-agers involved heavily with gangs in cities.[120] While urban slums were being blamed, a number of teen-agers from the suburbs could easily drive into a city and become involved with the turbulence there. Other "reasons" for the juvenile problems included rock and roll, more freedom, and lack of religious upbringing.

Philosophy

Critics of the progressive education movement had stated, by the 1950s, that schools had permitted less academic work than needed.[121] Therefore, there was a movement to go "back to basics" after progressive education in order to stress academics.[122] However, the latter part of this decade was dominated by the *Sputnik* crisis. Many Americans felt a "missile gap" between the United States and the Soviet Union.[123] While President Eisenhower was not as worried (probably because he had more information about America's defense than anyone else), the public and legislators became confident that Americans needed more emphasis on technology. Therefore, in 1958, Congress created the National Aeronautics and Space Administration (NASA) to coordinate research. Also in 1958, Congress enacted the National Defense Education Act which provided federal monies for training in languages, mathematics, and science.

More students wanted to go to college, and there were not traditionally enough colleges as colleges had seen such a decline in attendance during World War II. Therefore, more colleges and universities developed, and colleges began raising their standards for entrance as well as placing more emphasis on the sciences.[124] At the same time, schools dominated by minority or low income students did not provide adequate education to support their needs; instead, all effort was initially focused on very academically capable children—often from middle class families. This is also the time community colleges (mainly two year colleges) developed. This helped alleviate the large numbers of children (baby boomers) wanting to attend colleges and universities. All levels of education probably saw more government money through programs during the latter half of the 1950s and early half of the 1960s.

Conant continued his promotion of two tracks of students—those most capable for higher education and those who should be trained for a vocation. He began a study funded by the Carnegie Corporation in September 1957 that supported that a majority of students were being educated vocationally with a minority for further academic work within the same school successfully.[125] He felt this comprehensive school was doing well to educate America's children. His recommendations included closing smaller schools and that the academic tracked students needed to be challenged even more.

However, by the end of the 1950s, Americans were not happy with the standard of education as they felt that students did not have strong enough academics of standard subjects such as English and mathematics.[126]

Mathematics content

Near the end of "progressive education," very few of the school tracks offered in high schools included mathematics. The academic track did, and the business track and general curriculum track required little mathematics, but many students could graduate from high school having taken no high school level mathematics.[127] The beginning of this decade saw a general push towards traditional subjects as opponents came out against progressive education and the lack of academic rigor that may be needed for the Cold War.[128] There was an emphasis on traditional subjects of algebra and geometry for all students. Some students were even taking an advanced algebra or trigonometry course.[129] This switch from "life skills education" to a more academic one also crossed the beginning of a new reform movement begun during this decade: the "new math" movement.[130] One outcome of the "new math" movement was the request and documentation of some schools offering calculus—a course offering that is seen today because of the initiation of it at this decade.[131] This became heightened even more with the launching of *Sputnik* and the emphasis the general public would begin to place on math and science at the end of this decade. In fact, some say that the change in emphasis towards academics came overnight after *Sputnik*.[132] What is ironic, however, is that many of the teaching strategies that eventually were stressed with the "new math" movement to follow this outcry were more in line with progressive education than traditional education.

The "new math" movement consisted of many university, federally funded projects led by mathematicians and mathematics educators.[133] Mathematics curriculum and content became stronger during this time by stressing the need for understanding of mathematics and problem solving. The focus was on teaching arithmetic through set theory, incorporating mathematical language and symbol manipulation as well as studying more abstract concepts. The curriculum first became focused on college-bound students, but eventually filtered into mainstream mathematics for children of all ages. While there was not consistency in what was taught at each grade level because there were so many projects with different leaders and ideas, the common theme was that students were ill-prepared with mathematical knowledge, and mathematics needed to become a major focus for improvement in schools. Some of the topics that were stressed during this decade that had not been given much attention previously included: set theory, deductive methods, vector analysis, limits and functions, and probability and statistics.[134] Mathematicians became involved with most projects and stressed a need for higher levels of mathematics.[135]

Teacher education

Prior to the 1960s, teacher education for elementary teachers was fairly uniform with pre-service teachers taking one class in mathematics if it was required by the institution, and one "teaching arithmetic" course.[136] However, the CEEB (College Entrance Examination Board) in 1959 called for a vast improvement to teacher training in mathematics.[137] The CEEB stated that elementary teachers should come to college with a strong background in mathematics and continue taking mathematics while in college. It stated that high school X should take 24 hours beyond calculus with a minor in a related field. Other reports followed stating the importance of a strong mathematics background of high school teachers in fields such as probability and statistics, algebra, analysis, geometry, and applications of mathematics. Other groups went on to state that teachers should have various levels, and those teaching higher levels of mathematics in schools should have a master's degree in mathematics or a bachelor's degree in mathematics with a fifth year of teaching content (a suggestion that has evolved in some places in the United States today).

Pedagogy

The pedagogy of the teacher working with the child continued throughout this era—even as the new math movement became strong. There was a continued emphasis on "discovery learning" or the teacher as facilitator idea for children as they learn and understand things themselves. In fact, NCTM stated that the need for drill was becoming less important because of the advancement of machines (computers or calculators) to complete calculations. The emphasis turned towards thinking about mathematics and truly understanding it. "We must teach our students to do the work that machines cannot do."[138, p.424] Therefore, teaching for understanding became a focus of the "new math" movement of this decade.

Assessments

The 1950s saw more achievement testing with the advent of machine grading.[139] For the first time, a large number of students at the state level or national level could be tested and scored.[140] A complaint of this movement was that students were learning skills and algorithms in a passive manner in order to answer these same types of questions on the tests.

It was after this time that many elementary, as well as secondary students were tested each year. Also, the Committee on Advanced Placement of the College Entrance Examination Board (CEEB) assumed the responsibility for implementing more advanced coursework and assessments in 1955.[141] This was the beginning of the Advanced Placement (AP) tests that are prevalent today. The CEEB finalized its report in 1959 stating specifically what students needed to be able to do to enter college.

1960—1970

This is the era of the climax and then downfall of the "new math" movement. While students were exposed to new topics, new points of view (placing emphasis on meaning and understanding), and pushed down mathematics during the "new math" movement, many agreed that it was at the expense of basic skills and computation.[142] The 1960s was a time for excitement in the mathematical community, but teachers soon became disgruntled with a program that they did not help initiate; parents became frustrated at a curriculum they did not understand. Therefore, by the end of the 1960s, there was the beginning of a trend to head "back to the basics."

Historical context

This decade saw much social change as citizens protested for equal rights and civil rights. This included rights for African Americans, women, Latinos, and Native Americans.[143] The Cold War was still affecting Americans—in particular during the Cuban missile crisis—a few days that many Americans would say were the closest to a nuclear war with the formal Soviet Union. Kennedy was also involved in the troubles in Vietnam that were escalating at the time of his death.

With the assassinations of John F. Kennedy, Martin Luther King, Jr., and Robert Kennedy, the country found itself in a state of sadness—it appeared the country was torn between those trying to achieve equality and those angry at the changes. President Johnson worked to expand social welfare programs for many. His "Great Society" focused on social programs and a "War on Poverty." Some of the social reforms that were initiated under Johnson's direction included aid to poor children for schooling and housing, work-study jobs for college students, and Medicare and Medicaid—health insurance programs for elderly and poor, respectively. Johnson was also responsible for changing the immigration laws—the first time since the 1920s—where immigrants from all over the world would be allowed into the United States followed by their family members.[144] This allowed for fewer European immigrants and more immigrants from Asia, Korea, and Mexico than previously. Johnson also called for a movement towards progress with the Civil Rights movement in memory of President Kennedy—and the Civil Rights Act of 1964 was the first step towards government enforced equality.

While at the same time, a "counterculture" developed promoting a more provocative lifestyle than ever before. The "baby boomers" were attending colleges during the 1960s, and they had grown up in an era of prosperity rather than war. Many of these students participated in civil rights demonstrations while others worked with the Peace Corps and the VISTA (Volunteers in Service to American—the domestic version of the Peace Corps). Changes also included the wider-spread use of illegal drugs, a "hippy" lifestyle, and an increase of rock and roll music. However, the most focused aspect of this counterculture was the protests against the Vietnam War during this decade.

At the same time, the United States was pushing funding and emphasis in the "space race" with the first man walking on the moon in 1969—a date where many Americans felt they had "beat" the Russians in the space race. While Americans seemed to increase their technological skills during the space race, many Americans felt that the nation was experiencing a type of "cultural lag." This can be explained by the country advancing in some areas while neglecting basic needs in others—such as the problems with poverty within its own country.

Philosophy

The 1960s saw few required courses in schools and more electives—providing students with more choices became important. Since so many of these students were being admitted to colleges without a college preparatory background from high school, high schools continued to offer a wide variety of courses for students to take without emphasis on any particular academics.[145] Also, the Elementary and Secondary Education Act (ESEA) was passed in 1965 providing funds to schools to aid disadvantaged children—these funds have become known as Title I programs.[146] While students were encouraged to study more mathematics than they had before, schools were pressured to offer more academics than they had during the first half of the century. However, with the counter culture and the racial problems in the country, much of schooling was not focused on academics and was forced to focus on political and social issues.

Mathematics content

The "new math" era, which dominated the latter 1950s, 1960s, and briefly into the 1970s, was grounded in various projects throughout the United States. The "new math" movement dwelled on the conceptual understanding of mathematics rather than the computational understanding which had long been the focus for many teachers.[147] Teaching for meaning and understanding became a goal during this decade. Some of the ideas or changes brought about in the "new math" era that are seen today include: increased emphasis on geometry, probability, and statistics.[148] The Commission on Mathematics of the College Entrance Examination Board (CEEB) stated the following for college-bound students in mathematics: logic, statistics, and probability should be a part of school mathematics; plane and solid geometry should be integrated into one course; trigonometry should be taught with a second level algebra course; and pattern seeking should unify all mathematics.[149] Many of the "new math" projects and movements that produced materials for schools followed these recommendations for the new teaching materials. These recommendations are what led to what many high schools offer presently.

A variety of projects constituted the "new math" movement; with the most popular being SMSG—the School Mathematics Study Group. Most projects developed course materials around the CEEB recommendations led by mathematicians at universities and some mathematics educators and teachers. One of the largest complaints about the "new math" movement has been that it focused on the high school college bound when it also needed to focus on elementary and middle school programs. There were a few programs which did this (University of Maryland Mathematics Project for junior high and the Madison project for elementary), but the majority of the programs concentrated on high schools. Many of the projects had hoped teachers would use discovery learning in teaching the material, but few projects presented content in this manner.[150]

While the "new math" gained national attention and was present in many schools, the fact remains that in the entire United States school population, very few students were exposed to "new math." In fact, one researcher stated that, "It was possible at the time to walk into almost any school in the United States and see mathematics teaching that was little different from typical teaching before World War II."[151, p.625]

Teacher education

During the post-Sputnik era, teacher education came under fire for being too easy and not academically challenging. While most teachers were graduates of four year universities, and while most states were accountable for teacher certification, little had changed since the beginning of the century regarding the level of mathematics required of elementary teachers.[152] Only twelve states in 1960 required a specific course in mathematics, and this requirement was usually arithmetic or general mathematics. Thirty-five states required college graduation for certification. Many of these institutions, however, would admit elementary teacher candidates with no high school mathematics background. From the 1920s, the minimum number of mathematics credits required for graduation was raised from 24 to 27 for secondary teachers and 12 to 18 for elementary teachers. Therefore, while there was a gradual movement to increase the level of mathematical knowledge of teachers, it had taken some time and would need more time to raise the level even higher.

Conant and others called for teachers to take higher levels of academic courses in order to teach mathematics to academically strong students.[153] He also requested that teacher candidates have three years of high school mathematics and take at least six credit hours of college mathematics with teachers choosing mathematics as a main subject area to take at least thirty credit hours.[154] In fact, Conant suggested that schools provide many different courses, and that schools employ enough teachers to do so.[155] It was at this time that Conant also warned teachers and schools about issues they may soon face with students regarding crime, violence, and drug use.

Part of the problem with the "new math" movement was that the teachers were not adequately trained in the new materials—nor did anyone ever assess whether they had the background knowledge from their own education to teach this new material.[156] Thus, teacher education was overlooked, and many took for granted that teachers would just be able to teach these new materials developed by mathematicians during this reform era.[157] It became apparent to many that strong mathematical content would not solve the problem of improving mathematics in schools—mathematics instruction would also need improvement.[158]

The Committee on the Undergraduate Program in Mathematics (CUPM) published a report on teacher preparation in 1961 (this report was revised in 1964 and 1966).[159] The CUPM report recommended that elementary teachers take courses in number systems, algebra, and informal geometry as minimum requirements.

Pedagogy

The pedagogy most associated with the "new math" movement has been "discovery learning." While there are many words to describe the type of pedagogy used during this era, such as construct learning, "hands on" learning, etc., the basis of "discovery learning" is similar to those ideals found in progressive education. That is, students bring to a new problem experiences from the past, and then question and analyze an issue as they work towards a solution. The role of the teacher in the discovery classroom becomes more advisory and one of guidance than telling of factual information.[160] Teachers are still responsible for direction of content, concise working, and evaluation of the students.   However, discovery learning can be difficult for teachers to implement with little training in how to do so.[161] Incorrectly, many felt that discovery learning meant that students discovered everything about mathematics themselves. On the contrary, teachers expected to be involved, but students should be active learners by discovering patterns and suggesting methods for solving problems.[162]

Bruner's theory that any subject can be taught to any child at any age if the teaching was sound was discussed during this decade (Bruner also introduced Piaget's theories). Government funds "(Sputnik 'fear' money)" became available to support Bruner's theory in schools.[163,p.7] Gagne was another psychologist who studied children's understanding of mathematics, and he stated that children must transfer existing knowledge into problem solving techniques.[164]

This type of pedagogy was also used by the "open education movement" which followed the ideals of progressive education in that students directed their own learning with the facilitation of a teacher.[165] The "open education movement" was dominated by the successful sale of A.S. Neill's book *Summerhill* about a progressive school in England. Rising enrollments of students in Britain with a lack of classroom space (in particular in World War II bombed out buildings) forced British teachers to teach in more open spaces.[166] In fact, the open education movement, "was nothing new; it was just a repetition of progressivist programs promoted in the 1920s."[167, p.8]

Assessments

Standardized testing became popular after the passing of Title I of the Elementary and Secondary Education Act of 1965, as the federal government required school districts that received any funding of this sort to test students for evaluation purposes.[168]

Many of the "new mathematics" projects performed assessments comparing students taking the "new math" curriculum and those taking the standard curriculum. Many of these small scale studies showed that when comparing students of traditional curriculum with a "new math" curriculum, the students performed almost equally with those from the traditional curriculum doing better on calculations, and those from the "new math" curriculum performing better in problem solving skills.[169] However, many of the assessments failed to show that the goals and objectives of the "new math" programs were achieved (comprehension rather than only computation).[170]

One of the reasons that the next decade brought the "back to basics" movement was the continual failing of student performance on standardized assessments. Basically, student scores continued to decline through the 1960s into the 1970s. However, what has not been pointed out before is that student scores in *many* subject areas declined during this time (not only in mathematics); therefore, the validity of the tests as indicators of quality mathematics education should be questioned.

1970—1980

The 1970s era became known as the "Back to the Basics" era as computation and algorithms were stressed once again. Commercial publishers "quickly jumped on the 'back to basics' bandwagon."[171, p.349] Some national studies indicated that, by the mid 1970s, the majority of schools were using traditional textbooks with little "new math" influence.[172] It was during this decade that research in schools and mathematics education became focused and more popular.[173] This was also an era known as the "Decade of Accountability" as teacher success was based on student and school performance on standardized test scores.[174] Hence, a "dumbing down" of the curriculum took place.[175]

Historical context

The beginning of this decade was masked by the continuing Vietnam War—a war that many Americans did not back, and much of the first part of this decade focused on the US pulling out of Vietnam. Many returning soldiers had drug and alcohol problems, and some had severe psychological disorders.[176] The infamous Kent State shootings of students and many student protests caused even less support from the general public for the American government. In fact, many Americans began to feel that service to the military was dishonorable. Oil problems also led to high gasoline prices, and Americans began to see a more difficult economy than it had since World War II.[177] It was also at this time that many baby boomers sought work, and the workforce grew by 40 percent—as one may suspect, there were not enough jobs.[178]

The 1970s saw many Presidents as Lyndon Johnson finished his term, Richard Nixon resigned, was replaced by Gerald Ford, and then Jimmy Carter won the presidential election in 1976, to be beaten in 1980 by Ronald Reagan. President Johnson also began relations with both China and the Soviet Union by visiting those countries and meeting with leaders—a step towards more peace. However, the economy under Presidents Johnson, Ford and Carter continued to suffer.

Lifestyles changed in the 1970s as families had fewer children, and many women began working.[179] Families found it necessary to have two incomes with the increase in housing costs. With the economy in such bad shape, Americans were forced to work hard and deal with high interest rates and mortgage rates.[180]

Philosophy

The 1970s saw adults, citizens, and lawmakers questioning the role of teen-agers in society as they graduated from school and entered the workforce. The number of needed unskilled workers or blue-collar workers was gradually declining as factories closed and production was established outside of the United States. At the same time, more skilled workers were needed, and literacy became an issue at most jobs.[181] The 1970s saw students taking fewer academic courses, studying less, and learning less.[182] The schools backed down against student's rights, and at the same time, failed to make students responsible. The American public was concerned that schools were not educating students to become the types of productive citizens that would be needed as manufacturing slowed in this country and jobs requiring more education began to flourish. Teachers, forced to be held accountable for students with issues beyond their control, developed a "minimal standard" that students should attain describing the situation that occurred in schools above.[183]

Mathematics content

For most students graduating from high schools in the 1970s, they only needed to take one course in mathematics, and many of these students never took a course beyond the traditional one offered in ninth grade.[184] The 1970s saw a "back to basics" movement in mathematics as many Americans were not happy with the "new math" movement of the previous decade that seemed to produce children who were weaker in computational skills than they had hoped.[185] In fact, very few students actually saw the "new math" of the 1950s and 1960s, so when there was a call for "back to basics" most schools were back to teaching the curriculum and way they always had.[186] Students were seen capable in mathematics once they could master these basics which were defined as computational or pencil-and-paper skills. Another reason that the "back to basics" movement became popular was the drastic cuts in federal funding to the "new math" movements of the previous two decades. The focus became on basic arithmetic operations with little emphasis on problem solving or applications.[187] Words were taken out of textbooks and replaced by numerous mathematics exercises stressing the same content.[188]

While the majority of schools participated in a "back to basics" movement, mathematics educators were unhappy with this sudden switch away from teaching for understanding. In 1975, the National Advisory Committee on Mathematical Education (NACOME) studied the state of mathematics and offered suggestions in its published report as to the future direction of mathematics in schools. The NACOME report stated that the decline in standardized test scores could not be attributed to the "new math" movement, and that the "new math" movement had not reached as many schools and students as originally thought.[189] Recommendations from the report included finding a balanced curriculum for all students in mathematics.

Teacher education

The 1970s saw teacher accountability linked with student scores as a measure of teacher competence issues from the "top-down."[190] This is what drove the "back to basics" decade as many felt that poor teaching was leading to poorly educated citizens.[191]

Teacher education for elementary teachers during the 1970s typically consisted of one or two mathematics courses focused on number systems, some geometry, and mathematics methods.[192] The 1970s saw a call for more mathematics for teachers. However, while content and teaching are both considered important, research implies that an increased level in content automatically makes the teacher a better instructor; however, it is difficult to separate the two. In fact, "How an idea is represented is part of the idea, not merely its conveyance."[193, p.85] Therefore, while there may be a call for more content, the art or focus of teaching must be equally as important.

Pedagogy

The "open education" movement continued during this decade, and many schools constructed their classrooms in an "open" manner with no walls separating them.[194] In high schools, graduation requirements were reduced, offering more electives instead. Critics' responses to these classrooms were to go back to "basic skills" testing, and the curriculum moved away from child centered by the mid to late 1970s towards the "back-to-basics" movement. Most teaching was drill-and-practice based in direct-instruction.

It was also during this decade that the use of calculators and computers came into play in education. While they were still too expensive to have widespread classroom use, the implementation of calculators and computers was beginning.[195]

A common complaint of this era is that solving mathematics problems was a process taught by a teacher that students had to replicate providing one correct answer.[196] In other words, students had to remember which rule would work in a given situation rather than stressing the need for problem solving, analysis, and reasoning that would become a focus during the last quarter of the century. Furthermore, mathematics classrooms operated the same across the country starting the mathematics period by going over homework, then working some new problems or a new concept, and then starting a new set of homework problems in class.[197]

Assessments

The 1970s saw the start of the NAEP, the National Assessment of Educational Progress, a federally funded testing program that has continuously evaluated American students through the use of standardized testing.[198] Achievement tests remained popular in schools during this time—in particular because of the Title I law about assessment, and by this decade, almost 90% of schools received some type of Title I allotment—making testing each student the easiest method to show evaluation.[199] Accountability of teachers and schools became popular with the use of these tests; therefore, the curriculum forced popularity of a lower-level skills multiple-choice tests.[200] Standardized test scores were declining, and many attributed this to the "new math" movement—promoting a back to basics movement even more.[201] This caused an increase in standardized tests to see if students were mastering mathematics "basic skills." Another problem with standardized testing brought to the attention of others at this time was the fact that standardized tests had for so long been standardized against white, middle class students; thus, there was a bias quite possibly in the tests towards minorities.[202]

1980—1990

An Agenda for Action was published in 1980 by the National Council of Teachers of Mathematics (NCTM). This became a strong position statement placing NCTM at the head of mathematics education. It was in this document that NCTM emphasized the need for problem solving skills and the implementation of calculator use in the classroom. Another popular publication of the 1980s was *A Nation at Risk*, a report issued by the National Commission on Excellence in Education that stated that America should be worried about its state of education.

By the end of the 1980s, it was clear that technology would become more prevalent and necessary for students; thus, technology would become more of the focus in the 1990s. The end of the 1980s also saw the publication of the *Curriculum and Evaluation Standards for School Mathematics* by NCTM that focused on what children should learn in mathematics in grades K-12.

Historical context

This decade was known as a "Republican" decade with the leadership of Reagan and Bush. Ronald Reagan was elected in 1980, and his first term saw some hardship as he tried to sustain the economy while maintaining a large amount of government funds to national security and the Strategic Defense Initiative research program. Reagan continued to spend money—cutting programs that directly affected low income women and minorities.[203] The federal debt tripled from 1980 to 1989. The latter half of Reagan's presidency included meaningful and peaceful talks with Soviet leader Mikhail Gorbachev. At the same time, the USSR was deteriorating as Communist governments in Eastern European countries were collapsing.[204]

This decade was marked by many as the arrival of the information age. While the US had used computers in the 1960s and 1970s, it was during the 1980s that widespread use of computers came into homes and businesses. Many "blue-collar" workers were displaced as America began acquiring goods manufactured outside of the US. For the first time, Americans became skeptical of job loyalty and security. Baby boomers entered middle age, and called for more religious and moral values.[205] The 1980s also saw cities grow and rural America decline. Women continued to participate in the workforce at a higher percent than ever before. America saw more Hispanic and Asian immigrants than it had in the past. Homelessness increased substantially during the 1980s, and the AIDS epidemic began. Drug use was on the rise—in particular cocaine addiction.

Philosophy

The 1980s saw a shift away from electives to more emphasis on "core" subject matter including English, reading, science, mathematics and technology. Students were attending colleges in large numbers, but after their bad preparation in high school, many students had to spend time taking remediation courses in college. Therefore, schools began to analyze what was needed by students to succeed academically. It appeared that students were doing poorly not necessarily because of bad teaching, but because of a lack of a good and cohesive curriculum.[206] After an eighteen month study, *A Nation at Risk* concluded that secondary school curriculum was not educating students at a consistently high standard.[207] Its recommendations at the high school level included: four years of English, three years of mathematics, three years of science, three years of social studies, and half a year of computer science. These were considered "basics," and college preparatory courses should also be offered at an even higher level. Another recommendation was that foreign languages be started in elementary school, and that elementary schools focus on problem solving skills, science, social studies, and the arts in addition to English development. The report stated that there was a high demand for teachers to teach mathematics, languages, sciences, and the gifted and handicapped. At the end of the decade, President Bush initiated the goals that all students should achieve by the year 2000.

Therefore, schools began to stress academics more and vocational education less. Another different aspect of *A Nation at Risk*, was that it was the first report that stated that all students should be educated at their highest ability—not that students should necessarily be tracked.[208]

Mathematics content

At the beginning of the decade, most schools still only required one year of mathematics for graduation. A common complaint of the "back to basics" movement was that students needed a broader range of "basics" beyond algebraic and arithmetic computation.[209] Therefore, the National Council of Supervisors of Mathematics issued a statement in the late 1970s that framed the "basic skills" needed in the 1980s: these were to include computational skills as well as problem solving skills and applying mathematics. In 1980, NCTM published *An Agenda for Action* promoting similar changes in mathematics curriculum with a focus on problem solving.[210] This focus on problem solving was also to include the use of applications in mathematics and applications in how mathematics relates to other disciplines.[211] This report suggested that problem solving be needed with basic skills, and that calculators be brought into the curriculum.[212] Other issues included: basic skills defined as those beyond computation; calculator and computer use as appropriate; standards applied to teaching and assessments other than traditional; and more mathematics required of all students.[213] This report dictated that students in high school should take at least three years of mathematics with options to serve all levels of learners.

The definition of "basic skills" came into question as what was seen to be needed for the future that was considered basic included technology, problem solving, and computation. In fact, the report stated that basic skills should include the above as well as reading, interpreting, and constructing tables, charts and graphs; using estimation; understanding computers; applying mathematics in everyday life; and geometry and measurement. Also, the positive aspects of the "new math" movement were being recognized into the 1980s.[214]

Teacher education

A Nation at Risk stated that many teachers were not academically qualified to teach at the level needed in mathematics, science, and languages, and that more specialists were needed for education of the gifted and disabled. The report called for teachers to receive a higher level of academics in their preparation, and that their salaries rise to attract the best and brightest. Most teachers were taking courses in methods but not in their content area; and many teachers who taught mathematics and science were not qualified to do so.[215] Regarding mathematics specifically, most teachers were taking basic level courses or courses designed for teachers only. The Holmes Report during the 1980s began its "Agenda for Improving a Profession" comparing the lack of teaching seen as a profession when compared to other professions.[216] One agenda that evolved from this time period was the idea of standards that all teachers should meet in their preparation and their teaching positions.

Teacher preparation again called for a higher level of mathematics taken at the college level for secondary school teachers. Now, NCTM was recommending that teachers take courses in the history of mathematics, topology, differential equations, linear algebra, graph theory, logic, computer science, abstract algebra, and/or combinatorial analysis.[217] Elementary teachers were also called to take more mathematics at the college level to include discrete mathematics, algebra, geometry, and possibly an introduction to calculus. There was also again a call to help in-service teachers by providing professional development.

It was during this decade that the National Council of Teachers of Mathematics became focused on creating standards for school mathematics. These standards were published in 1989 (*Curriculum and Evaluation Standards for School Mathematics*), and became the framework for school mathematics in the next decade.[218] This framework centered on themes of mathematics such as problem solving, communication, reasoning, and mathematical connections. It also stated what mathematics content topics should be taught in appropriate grade levels K-12. The document stressed the need for problem solving within a balanced curriculum stressing applications and connections.

Pedagogy

In the same report listed above from the National Council of Supervisors of Mathematics (NCSM), it suggested teaching pedagogy to encompass more teaching methods rather than just drill-and-practice with less emphasis focused on standardized testing. In light of both *An Agenda for Action* and *A Nation at Risk*, mathematics education became a focus in schools with the promotion of understanding mathematics and problem solving for all levels of students— setting a higher standard for all students than had been set before.[219] To be able to teach children to truly understand mathematics, they would need to construct the knowledge themselves rather than constantly be taught in a direct-instruction manner.

The National Research Council published *Everybody Counts* at the end of the decade, and one of its components stated that children must make sense of their own mathematical learning. It stated that research in learning showed that most students actually construct their own understanding based on new experiences that broadens the framework in which ideas are shaped.[220] This fits exactly with the framework of pedagogy stressed in Dewey's theory of progressive education. The report continues that teachers should be facilitators of learning rather than continue to dwell in the traditional lecture style of mathematics teaching.

The 1980s also saw a movement for more research in mathematics education beginning with the reflective practice of teachers who could reflect on their teaching and mathematics to serve students best.[221]

Assessments

Assessments given to students became a highlight of national concern with *A Nation at Risk* and other reports that showed low test results of American students. Therefore, tests were used mainly for policy reformation without fair analysis of the students. NCTM also called for more evaluative methods beyond the standardized test which had been the norm for so long.[222] American students are the most tested students in the world.[223]

Three years after the publication of *A Nation at Risk*, an April 1986 report found that forty-one states had increased their high school graduation requirements, thirty-three states developed competency tests, thirty states initiated teacher competency tests, and twenty-four states had started teacher salary enhancement programs.[224] Therefore, standards and assessments of both students and teachers were raised. A complaint of this effect is that despite raising standards, the effects of parents, home life, and health on children's performances were not addressed.

1990—2000

With the *Standards* leading the way in the 1990s, many educators began to focus on different teaching styles involving collaboration, communication among students, cooperative learning, etc. Also, many states and school districts began developing curriculum guides that were based on these standards. In fact,

> With the exception of the launching of Sputnik in 1957, no other single event seems to have sparked as much public controversy and debate in school mathematics as the advent of the standards movement.[225, p.282]

This is also the decade that initiated the "math wars" between those who favor problem solving and those who favor drill and practice (the dichotomy described in the introduction). Critics felt that we were headed for a new, new math or "whole math."[226] Technology became much more prevalent in the curriculum as calculators and graphing calculators became standard in many classrooms. This decade ends with the publication of *Principles and Standards for School Mathematics* by NCTM in 2000 focusing again on mathematics learning for all.

Historical context

This decade saw the dissolution of the Soviet Union, the end of the Cold War (and the tearing down of the Berlin Wall in Germany), more taxes as medical costs and government spending rose, and the Gulf War where Americans were deployed in the Middle East. The 1980s had also seen much drug abuse in its citizens, and President Bush declared the "War on Drugs" in the 1990s. With the quick end to the Gulf War, the election of President Clinton in 1992, and the continuing expanse of technology and computers, many Americans felt prosperity during this decade.

This decade saw a huge increase in the number of personal computers, video recorders, CD and DVD players, mobile cell phones, and the use of the

Internet and e-mail as means of resources and communication.   Women continued to enter the workforce, and the "family unit" continued to decline as many people lived independently. Many people were postponing marriage into their 20s and 30s; also, the divorce rate rose. Immigrants continued to arrive in droves, and the majority came from Asia, Mexico, and Latin America. Some of President Clinton's reforms included the Family Medical Leave Act, student loans, and an increase in the minimum wage.[227] On the other hand, drug use and extreme violence continued to disrupt young people's lives—particularly with gang violence in inner cities.[228]

The end of the decade saw the election of George W. Bush who would pass the "No Child Left Behind Act" in May 2001, and then face one of the toughest first years in office with the September 11, 2001 attack on the World Trade Center.

Philosophy

One of the largest efforts towards education was initiated by the Bush administration and continued by the Clinton administration: Goals 2000. This included a set of goals for all American children to be achieved by the year 2000. They were:

1. All children in America will start school ready to learn.
2. The high school graduation rate will increase to at least 90 percent.
3. All students will leave grades 4, 8, and 12 having demonstrated competency over challenging subject matter including English, mathematics, science, foreign languages, civics an government, economics, the arts, history, and geography, and every school in America will ensure that all students learn to use their minds well, so they may be prepared for responsible citizenship, further learning, and productive employment in our nation's modern economy.
4. United States students will be first in the world in mathematics and science achievement.
5. Every adult American will be literate and will possess the knowledge and skills necessary to compete in a global economy and exercise the rights and responsibilities of citizenship.
6. Every school in the United States will be free of drugs, violence, and the unauthorized presence of firearms and alcohol and will offer a disciplined environment conducive to learning.
7. The nation's teaching force will have access to programs for the continued improvement of their professional skills and the opportunity to acquire the knowledge and skills needed to instruct and prepare all American students for the next century.
8. Every school will promote partnerships that will increase parental involvement and participation in promoting the social, emotional, and academic growth of children.[229]

One of the main goals of this act was to help improve teachers' skills, increase parent involvement, and encourage more focus on meeting the needs of children in this country. In general, however, the American public is often slow to acknowledge necessary change; and the goals and standards movement is no exception. While policy makers and others advocate for change, there are many who do not support it in particular if funding is seen as a problem.

Another aspect of both the 1980s and 1990s was the decline of vocational education in the high school and the increase of it in the community college. Also, this has become known as the Standards Based era as government and the public are asking for accountability and measurement on education. While most states have standards and many agree that some type of standards-based reform is necessary, it remains to be seen how the standards reform will progress.

Mathematics content

The 1990s saw the biggest development of standards in mathematics compared to any other field. It was during this decade that the National Council of Teachers of Mathematics *Standards* (that were published in 1989) became the focus of school mathematics. By 1997, most states had developed mathematics standards that were in line with those put forth by NCTM.[230] NCTM called for understanding of mathematics, the development of critical thinking and problem solving skills, and stressing fewer computation skills than had been the norm of the 1970s back to basics movement. Students were also expected to use calculators often now that they were widely, and cheaply, available.

Because of the widespread use of the *Standards*, many state and local districts developed standards and curriculum for mathematics around the publication. What entailed in the 1990s became known as the "math wars" between groups who favored basic skills and those who favored a problem solving approach. The mathematics taught in this decade has been called by some critics as the "new New Math" or "fuzzy math."[231] Again, the question of how much emphasis should be on basic skills became a debated topic. Critics claimed that materials did not support fundamental algebraic or arithmetic skills.[232]

As this decade closed, the National Council of Teachers of Mathematics set to revise the standards and published *Principles and Standards for School Mathematics* in 2000. In this document, NCTM stressed basic skills and computational skills more than it had in the previous *Standards*.[233] However, the focus of the document continued to be educating all students to a high standard in mathematics involving the use of basic skills as well as problem solving. Debates about where mathematics education is headed as the decade closes is similar to the same debates that occurred during the 1950s and 1960s and the "new math" movement. There is constant debate as to where mathematics education should lead including the purpose and content.

Mathematics continues to be taught in high schools in separate courses of algebra, geometry, and so on. The United States is one of the few countries that does not teach unified mathematics throughout schooling.[234] Furthermore, there is constant debate as to the use of calculators in the classroom compared with the constant drill and practice of algorithmic computation that calculators perform easily today.[235]

Teacher education

While teachers today earn a minimum of a bachelor's degree in teaching, and many have a master's degree, the reality of where these teachers are and who they serve remains the problem. An issue with teacher training in the 1990s has been the lack of qualified teachers needed for all public schools. While most schools are served by adequately trained teachers and highly trained teachers, there is concern that inner-city or poverty stricken schools do not have teachers teaching subjects in which they are qualified.[236] This is extremely important in mathematics teaching above the grade school level where mathematics content can become more complex.

In particular, there is a question on the abilities of middle school and high school teachers to teach mathematics when they do not hold a mathematics degree themselves.[237] In fact, a major problem in mathematics teaching is the number of teachers who are not qualified to teach mathematics (they are instead qualified for another subject), but are still teaching mathematics. One study found that one-third of all high school teachers teaching a mathematics course do not have a major or minor in mathematics from college. Furthermore, almost one-half of middle school teachers teaching mathematics do not have a degree in that field. The National Science Board's Science and Engineering Indicators 2004 states that 18% of American high school students and 57% of American middle school students were taught mathematics by a teacher who did not have a major or minor in mathematics or a mathematics-related field.[238] What is also evident is that the majority of teachers teaching mathematics without a degree in mathematics is occurring in lower socio economic schools—causing the problem that the most disadvantaged students are being taught by the least qualified teachers.[239]

One of the biggest criticisms of teachers regarding this reform is that, as has been in the past, teachers were consulted least on how to improve schools when they are the ones closest to the issues.[240] Therefore, NCTM used teachers to help write the publication, *Professional Standards for Teaching Mathematics,* published in 1991 which outlined standards for teaching.

Pedagogy

The pedagogy that came with the NCTM *Standards* publication included a more student centered classroom with students learning materials with manipulatives and group learning.[241] The mid-1990s saw a revolt against the *Standards* by the Mathematically Correct group in California that stated that students needed more basic skills than NCTM had espoused.[242] Nevertheless, the pedagogy associated with the *Standards* was the same as the progressive movement—that the child-centered classroom would provide the best learning opportunity for the child. The theories of the 1980s were similar in pedagogical thinking of the 1920s and 1930s; they were called "constructivism" rather than progressivism.[243] While constructivism has been seen as an extreme form of learning where children construct all knowledge themselves, it is evident that the teacher would have to lead the children somewhat in their knowledge base.[244]

> Constructivism as a theory of learning can provide the framework needed to help math teachers move from a transmission model to one in which the learner and the teacher work together to solve problems, engage in inquiry, and construct knowledge.[245, p.521]

In fact, many would argue that constructivism in practice could follow the three stages of representation of mathematical ideas as expressed by Bruner in the 1960s: "enactive, iconic, and symbolic."[246, p.85] The first stage requires students to use manipulatives and hands-on materials to solve problems; the second stage moves students toward using pictures, drawings, or representation of some sort on paper; and the last stage then moves students into the abstractness of mathematics solving problems without other assistance.

One of the most important aspects of teaching not mentioned yet is the attitude of the teacher towards mathematics. It is difficult for teachers who were taught or had a negative experience with mathematics themselves as children to enlighten their students in a new way of teaching. Because they lack mathematical confidence themselves, their pedagogy is automatically affected.[247] Therefore, new teachers must be mentored and encouraged to find a method of teaching mathematics that enables students' abilities and attitudes towards mathematics (realizing that this may mean doing something to build confidence themselves in mathematics).

Assessments

NCTM also published, in 1995, *Assessment Standards for School Mathematics* which stated the various types of assessments (other than just standardized testing) that teachers and schools should use to assess mathematical ability. There is also a focus on equity in that assessments should avoid cultural bias and deemphasize traditional assessments as the main means of assessment.[248] However, quite possibly more now than ever before, America is seeing emphasis (mainly by the government) placed on assessments of children in order to "prove" that goals and objectives in their education are being met. This has become known as "high stakes testing," and has moved to the forefront of education since the beginning of the 21st century. In fact, "high stakes testing" has often become aligned with the Standards movement whether the believers of the Standards movement agree.[249]

What is known about testing today is that tests can only measure what they were established to measure; therefore, if a test is based on basic skills then the outcome will only be on basic skills—not necessarily the ability to problem solve as well. Therefore, in order to make tests more valid, the purpose of the testing and what exactly the test is set to measure must be clear.[250] There has also been a complaint that traditional American standardized tests are filled with short mathematics problems that can be solved quickly rather than more in-depth problems that would require higher order thinking skills.[251]

A concluding word about assessment is to bring to the reader's attention the fact that American students have participated in three major international tests, among others. These include the First (mid 1960s), Second (early 1980s), and Third (mid-1990s) International Mathematics and Science Study. Also, the National Assessment of Educational Progress (NAEP) has been analyzing student assessment in mathematics regularly since 1964. Data about student progress in mathematics can be found from these (and other studies). The only word of caution in comparing to other nations' children is to make sure that the same level of ability is being assessed by children who are similar in intellect and upbringing.

Conclusions

The past century can also be divided as follows: the time period from 1900-1930 was dominated by increasing numbers of students attending schools. In order to provide for all students, a vocational program was created alongside an already existing academic program in mathematics.[252] The time period from 1930-1960 was dominated mainly by more students attending high schools. The mathematics focus was on arithmetic and basic mathematics needs because it was assumed they would not need higher level mathematics in their lives; therefore, the majority of students would not need to be mathematically challenged. The time period from 1960-1980 was characterized by allowing much more choice for students in their graduation requirements. And the last time period from 1980-present day was the beginning of a focus on higher levels of mathematics for graduation requirements with less choice and more focus on achievement and standards. This could be related to the American society changing from first an agricultural society to an industrial society, and then more recently, from an industrial society to an information age society.[253] At each turning point, it was realized by many that students would be required to perform higher order thinking and solving in mathematics to obtain jobs in the new American economy.

Reformers tend to name a reform such as a progressive reform that spanned three decades, or the standards reform that has spanned almost two decades. While these names imply uniform movements from all schools; in reality this rarely happens. Thus, while some schools did participate, there were always a number who maintained a traditional course.

As one looks back across the last century, it is obvious that school reform does swing back and forth on the pendulum described at the beginning of this book. First, Progressive Education from 1900-1950 saw professional leaders re-examining schools.[254] Progressive education brought several new components that are a concern for schools today as well: health, family and community issues, and vocational education.[255] In particular, progressive education began the initial look at educating a more diverse group of students based on their interests and needs.

In fact, some of the pedagogical techniques used today stem from the "child-centered" era of the progressive movement. For example, inquiry-based learning, project-based learning, authentic learning, or problem based learning are techniques that found their roots in progressive education. Working together in groups is another technique that was popular during the progressive era. Drill and practice as a theory has always been debated. Even in the *1923 Report,* careful consideration is given to drill and practice as a regular means of learning with children not clearly comprehending the mathematics behind the practice. As mathematics education has swung on the pendulum, educators go back and forth between progressive ideals and drill and practice.

Ideas from the progressive era continued until "student-centered" came under attack as critics felt the current content of subjects was inferior to strong academics of mathematics and science that needed to be taught. However, after the launching of *Sputnik,* some of the same ideals of the progressive era can be found in the "new math" era. The "new math" era stressed that students should learn to solve problems—many attempted this through "discovery learning" which is providing an atmosphere where students can actively engage in learning mathematics.[256] The Standards movement also had these same ideals. In fact, Klein states that, "The NCTM *Standards* reinforced the general themes of progressive education, dating back to the 1920s, by advocating student centered, discovery learning."[257, p.13]

> The NCTM standards were a brew of progressivism—a nod to the 1920s when math was supposed to be practical—and constructivism, which was progressivism that adapted research from cognitive psychology to the task of teaching and called it discovery learning.[258, p.33]

One of the main problems with changes in any reform is that teachers, parents, and administrators must be in support of the reform. Too often in the past, reforms have not been as successful as hoped because new programs were chosen for teachers, and teachers and parents were expected to utilize and learn the new materials without much input.[259] The "new math" movement began addressing this issue when it required teachers to go to workshops to learn the "new math;" however, for reform to be successful, stakeholders must see and agree with the positive outcomes of the reform. The NCTM Standards did include more stakeholders in their 2000 publication—including teachers, mathematicians, mathematics educators, and others to look over drafts of the document before expecting all to blindly accept it.[260] This may be a "lesson" from the "new math" movement that was learned—to involve those closest to the teaching material as much as possible from the start, and to provide professional development opportunities for these teachers.[261]

While the progressive reform, "new math" reform, and Standards reform all had similar ideals in pedagogy and the child-centered classroom, what in effect happened in the United States, is that many classrooms never changed to follow these trends. Some teachers modified their teaching slightly, but a majority of classrooms really never saw a drastic change from the teacher directed classroom. Elementary classrooms changed more than high school classrooms in the progressive era, and high school classrooms changed more in the "new math" era, while both are being affected in the Standards era. Nevertheless, a look at the past decade has formed a general consensus for the mathematics curriculum in this country despite any formal national curriculum: a focus on arithmetic, fractions, decimals with some emphasis on descriptive statistics, geometry, and measurement for grades K-8, and a year of geometry, two years of algebra, with some options for calculus or the study of functions in high school.[262]

Connections

As there are a variety of connections between various reform movements and between different issues affecting mathematics, the connections of some key concepts are briefly described followed by connections to present reform.

Textbooks: One issue that has dominated the curriculum and the mathematics taught in schools throughout the century of reforms is that of textbooks. While reformers may have ideas of where mathematics education should change, the fact remains that little change is ever done without the advancement of textbooks. Because textbooks continue to drive curriculum frameworks for states and schools, change typically does not happen locally unless textbooks are available in that area. Thus, while reformers have ideas, change only takes place when curriculum is placed in the hands of the teachers or textbook writers who affect teachers.

Since 1900, almost half of the states have used some type of textbook adoption methods—this fact affects those states directly and all other states indirectly by limiting the types of materials available since a few authors or publishers work hard to get these states to "adopt" their textbook.[263] The problem with the "textbook-driven curriculum" is that it directly involves politicians with mathematics curriculum choices. It also allows for the swing on the pendulum from "new" mathematics to "old" mathematics to continue as states and textbook writers move back and forth. This is particularly true with the impact that the two largest adoption states have: Texas and California. As these two states choose curriculum, it indirectly affects the rest of the country because of the emphasis, money, and time that publishers and authors must spend in trying to get their textbook adopted in order to make money. This indirectly means that a majority of curriculum designed textbooks follow the "trends" of curriculum at the time of those in Texas and California. Therefore, while California has recently gone more "back to the basics" the pendulum swing continues for books written for that state compared to "standards-reform" books written for other states.

Unified mathematics: The idea of a unified mathematics course for all grades continued to be a recommendation throughout various reforms in the last century. Unified mathematics is established throughout the elementary and basic junior high or middle school curriculum, but very few high schools or higher level mathematics courses in middle schools use a unified curriculum. In the early 1900s, unification of mathematics also included science and mathematics taught in conjunction showing the connections between both.[264] Vocational courses of the first two decades of the twentieth century were unified in nature; however, the academic/college track courses continued to not be unified.

Teacher education: One major issue that comes to light in any reform is that of teacher education. One thing remains true: in every reform, teachers are called to know more about mathematics, curriculum, teaching, student understanding, and classroom management. There is a call to continually strengthen teacher education programs and in-service programs for teachers. There has been a constant need for better-qualified teachers.

Education of all students: What many reforms have done, without necessarily focusing their attention, is engage in the debate over who to teach or who education is for. While the turn of the century saw mainly college bound students only, the end of the century saw mathematics education focused for all students. Throughout the century, more students have enrolled in school and the importance of educational of all students has gradually developed over the century.

Technology needs: In the event of every reform movement, technology has increasingly become part of the mathematics taught in schools. From slate boards to paper to manipulatives to calculators and computers, technology has forced the teaching of mathematics to be constantly changed and updated.

Assessments: Assessments can be split roughly into two periods. The first fifty years of the twentieth century saw assessments as a means to controlling educational opportunity as appropriate. Assessments were used often as a policy tool for reform.[265] It was thought that assessments could be used to create a socially desirable community of learners and citizens. This was particularly important during the Progressive era. The next period is the last fifty years of the twentieth century where testing was used as a springboard for various mathematical reforms and explanation for dissatisfaction with mathematics education. This period included federal funding for testing to assess curriculum and school quality. This remains an important and controversial topic today.

Reforms: Regarding any reform movements, what remains constant is,

> That teaching is a very complicated, complex activity. Instructional decisions about grouping students; designing problems and tasks; and presenting appropriate, worthwhile mathematics are determined on the basis of teachers' knowledge of the learning process, learners' needs and interests, and a firm understanding of the mathematics to be taught.[266, p282]

In essence, one must be highly educated in mathematics with strong pedagogy. One is not more important than the other—a teacher must be equally good at both in order to provide the best mathematics instruction.

Mathematicians and mathematics educators have constantly called for more mathematics for students and teachers where the general public have not always strongly and quickly followed suit.   A major reform effort includes the following: profession support, public support, good curriculum materials, teacher education, understanding of clear goals of the reform, assessments that reflect the curriculum, and information about reform research results.[267] A reform without some of these components may have difficulty anywhere. This is why we see similar reform statements made by the mathematical community over time without quick and constant change actually happening in the schools.

Conclusion:

Looking back through the century, some of the reforms can include: Progressivism and the project methods of the 1920s; "life adjustment" courses of the 1940s; the "new math" movements of the 1950s; the "open classroom" reform of the late 1960s and early 1970s; and the Standards movement of the 1990s. All of these have some of their roots based in progressive education as the pendulum continues to swing back and forth. In fact, "The NCTM reforms are the latest manifestation of progressive education."[268, p.203] The goals of all these reforms continue to be the same: students should understand and value mathematics, communicate and reason with confidence in their ability to solve problems.[269]

It was thought that progressive education just "died out" as did the "new math" movement of the late 1950s and 1960s.   What can be seen when comparing the various reform movements throughout the century, is that education has consistently advocated a progressive education agenda for most of the 20th century.[270] Pedagogy or philosophy of education tends to shift together—despite the social changes of citizens and the country. For example, even though vocational education became popular during the Progressive Era because of the need for blue-collar workers at that time in our history, the pedagogy and philosophies of progressive education and standards based education are similar.   In all cases, progressive education, the "new math" movement, and the Standards movement all maintained that they were "the new or modern" education.

Furthermore, "new math" called for a "meaning-centered" program and the Standards call for a "problem-based" program, just as the progressives did at the beginning of the century. In all cases, computational skills were de-emphasized while problem solving skills were emphasized.[271] In this constructivist teaching pedagogy, students learn best by taking responsibility for their own learning.

However, is progressive education or standards education really teaching all students all the time using their interests as personal motivation? In reality, teachers must strike a balance because it would be impossible for all teachers to teach exactly the same way using the ideals of one movement.

Therefore, what is really needed is a balance of problem solving skills and basic computational skills—with both being taught in a child-centered classroom. There is no reason that teachers, parents, or students should have to choose between the dichotomies initially explained in this document:

the old and the new in mathematics
skills and concepts
the concrete and the abstract
intuition and formalism
structure and problem-solving
induction and deduction [272, p.136]
facts vs. higher order thinking
problem solving vs. basic skills [273, p.1]
radical reform vs. back-to-basics programs [274]

In fact, "it is an oversimplification to state educational issues as dichotomies, even though it helps to focus the debates that swirl around them."[275, p.456] Also, "...dichotomies in education are artificial and incomplete statements of problems."[276, p.457] Rather, mathematics programs should contain a nice balance between each pair, with emphasis between the two determined by the teacher and objectives of a lesson. Skills and conceptual understanding are intertwined, and a successful student is one who can do both. It would seem best that students be able to solve mathematical problems with computation skills and problem solving skills and mathematical reasoning. Furthermore, research shows that students cannot learn just by watching and imitating, but they must become actively involved in learning mathematics.[277] One type of constructivism that may utilize both aspects is *dialectical constructivism*—the idea that students can acquire knowledge through their experiences in some cases when they are teacher-led, and in some cases where they are student-led.[278] Furthermore, the idea of cooperative learning or students working in groups can also be used when the classroom is student-centered or teacher-led.

> Notwithstanding the NACOME verdict that the new math reform agenda was "fundamentally sound" and data suggesting that it had some lasting positive effects, the public memory of the new math period in mathematics education remained an image of mistakes and failures. That lasting impression reflects the American tradition of enthusiastic embrace and ultimate rejection of reform.[279, p.535]

Another important aspect for teachers, parents, and citizens to understand is that not all reform should be initiated and abandoned at such speed that no one learns the best characteristics of it as seen in the "new math" movement described above. In traditional reform, a sequence of events often takes place: some problem develops; school reforms are introduced to alleviate the problem; compromise is established that eases the problem; some type of national crisis or economic issues occurs; another type of school reform is introduced ending in another type of compromise, and the cycle continues.[280] Thus, it is imperative that educators begin to look to the past and understand that many reforms can teach us what to leave behind and look forward.

Furthermore, reform has often swung in contradictory directions of either equalizing educational opportunity or focusing only on higher academic achievement. Why should any reform have to choose one or the other...again a nice balance is necessary where teachers can provide a high level of mathematics for all students regardless of academic level. Most reforms grew out of a discontent with student mathematics performance or ability, and succeeded in furthering mathematics as a subject taught in K-12 schools.[281]

Jones and Coxford have stated that for many decades, the focus on mathematics education should be the following questions:

> What are the goals of mathematical instruction?
> What should we teach at each grade level?
> How can we organize our content and classroom methods to achieve these goals in a rapidly growing and diverse school population?[282, p.459]

If the goal is educating all to their highest ability in mathematics in an effective manner, then continually analyzing the above questions can strengthen mathematics teaching and learning.

While it is often difficult to deliver and sustain a reform movement of any kind, one of the most important concepts that arises from debates and history of mathematics education reform is the fact that the importance of mathematics education to America's children is brought to the forefront of many. Changes should constantly be made to enhance education, and these changes must be allowed time to occur with careful evaluation of success or failure. "Feeling 'at home' with mathematics is every child's right."[283, p.200]

Notes

1. Wu, H. 1996. The mathematician and the mathematics education reform. *Notices of the AMS* 43 (12):1531-1537.
2. Jones, Eric D., and W. Thomas Southern. 2003. Balancing perspectives on mathematics instruction. *Focus on Exceptional Children* 35 (9):1-15.
3. Stein, S.K. 1996. *Strength in numbers: Discovering the joy and power of mathematics in everyday life.* New York: John Wiley & Sons, Inc.
4. NACOME, National Advisory Committee on Mathematical Education. 1975. Overview and analysis of school mathematics, grades K-12. Reston, VA.
5. Loveless, Tom, ed. 2001. *The great curriculum debate.* Washington, DC: Brookings Institution Press.
6. Ewing, John. 1996. Mathematics: A century ago -- a century from now. *Notices of the AMS* 43 (6):663-672.
7. Fey, James T. 1978. The United States experience with New Math. *Educational Studies in Mathematics* 9 (3):339-353.
8. Shirley, Lawrence. 2005. Notes on history of math teaching and math books.
9. Parker, Franklin. 1993. Turning points: Books and reports that reflected and shaped U.S. education, 1749-1990s, edited by ED369695: U.S. Department of Education.
10. Garraty, John A., and Mark C. Carnes. 2000. *The American nation: A history of the United States since 1865.* New York: Addison Wesley Longman, Inc.
11. Burrill, Gail. 2001. Mathematics education: the future and the past create a context for today's issues. In *The great curriculum debate*, edited by T. Loveless. Washington, DC: Brookings Institution Press.
12. State, United States Department of. 2006. *Outline of US history* [Internet] 2005 [cited January 18 2006]. Available from http://usinfo.state.gov/products/pubs/histryotln.
13. Garraty, John A., and Mark C. Carnes. 2000. *The American nation: A history of the United States since 1865.* New York: Addison Wesley Longman, Inc.
14. Berube, Maurice R. 1994. *American school reform: Progressive, equity, and excellence movements, 1883-1993.* Westport, CT: Greenwood Press.
15. Tozer, Steven E., Paul C. Violas, and Guy Senese. 1998. *School and society: Historical and contemporary perspectives.* Boston: McGraw Hill.
16. Angus, David L., and Jeffrey E. Mirel. 2003. Mathematics enrollments and the development of the high school in the United States, 1910-1994. In *A*

history of school mathematics, edited by G. M. A. Stanic and J. Kilpatrick. Reston, VA: National Council of Teachers of Mathematics.

17.   NEA. 1970. Report of the Committee of Fifteen on Elementary Education. In *Readings in the History of Mathematics Education*, edited by J. K. Bidwell and R. G. Clason. Reston, VA: NCTM.

18.   Reys, Robert E. 2001. Curricular controversy in the math wars: A battle without winners. *Phi Delta Kappan* 83 (3):255-258.

19.   Grouws, Douglas A., and Kristin J. Cebulla. 2000. Elementary and middle school mathematics at the crossroads. In *American education: Yesterday, today, and tomorrow*, edited by T. L. Good. Chicago: The University Press of Chicago.

20.   Burrill, Gail. 2001. Mathematics education: the future and the past create a context for today's issues. In *The great curriculum debate*, edited by T. Loveless. Washington, DC: Brookings Institution Press.

21.   Good, Thomas L., ed. 2000. *American education: Yesterday, today, and tomorrow*. Chicago: The University of Chicago Press.

22.   Jones, Phillip S., and Arthur F. Coxford, eds. 1970. *A history of mathematics education in the United States and Canada*. Vol. 32nd Yearbook. Reston, VA: National Council of Teachers of Mathematics.

23.   Donoghue, Eileen F. 2003. The emergence of a profession: Mathematics education in the United States: 1890-1920. In *A history of school mathematics*, edited by G. M. A. Stanic and J. Kilpatrick. Reston, VA: National Council of Teachers of Mathematics.

24.   Rury, John L. 2005. *Education and social change: Themes in the history of American schooling*. 2nd ed. Mahwah, NJ: Lawrence Erlbaum Associates.

25.   Kilpatrick, Jeremy. 1992. A history of research in mathematics education. In *Handbook of Research on Mathematics Teaching and Learning*, edited by D. A. Grouws. New York: Macmillan.

26.   Parker, Franklin. 1993. Turning points: Books and reports that reflected and shaped U.S. education, 1749-1990s, edited by ED369695: U.S. Department of Education.

27.   Oliver, J. Steve, and B. Kim Nichols. 1999. Early Days. *School Science and Mathematics* 99 (2):102-105.

28.   Parker, Franklin. 1993. Turning points: Books and reports that reflected and shaped U.S. education, 1749-1990s, edited by ED369695: U.S. Department of Education.

29.   Garraty, John A., and Mark C. Carnes. 2000. *The American nation: A history of the United States since 1865*. New York: Addison Wesley Longman, Inc.

30.   Current, Richard N., T. Harry Williams, Frank Freidel, and W. Elliot Brownlee. 1986. *The essentials of American history since 1865*. New York: Alfred A. Knopf, Inc.

31.   Ravitch, Diane. 2000. *Left back: A century of failed school reforms*. New York: Simon & Schuster.

32.   Tozer, Steven E., Paul C. Violas, and Guy Senese. 1998. *School and society: Historical and contemporary perspectives*. Boston: McGraw Hill.

33.   Garraty, John A., and Mark C. Carnes. 2000. *The American nation: A history of the United States since 1865*. New York: Addison Wesley Longman, Inc.

34.   NACOME, National Advisory Committee on Mathematical Education. 1975. Overview and analysis of school mathematics, grades K-12. Reston, VA.

35.   NCTM, ed. 1926. *General survey of progress during the last twenty-five years*: National Council of Teachers of Mathematics.

36. Jones, Phillip S., and Arthur F. Coxford, eds. 1970. *A history of mathematics education in the United States and Canada*. Vol. 32nd Yearbook. Reston, VA: National Council of Teachers of Mathematics.

37. Grouws, Douglas A., and Kristin J. Cebulla. 2000. Elementary and middle school mathematics at the crossroads. In *American education: Yesterday, today, and tomorrow*, edited by T. L. Good. Chicago: The University Press of Chicago.

38. Kliebard, Herbert M., and Barry M. Franklin. 2003. The ascendance of practical and vocational mathematics, 1893-1945: Academic mathematics under siege. In *A history of school mathematics*, edited by G. M. A. Stanic and J. Kilpatrick. Reston, VA: National Council of Teachers of Mathematics.

39. Jones, Phillip S., and Arthur F. Coxford, eds. 1970. *A history of mathematics education in the United States and Canada*. Vol. 32nd Yearbook. Reston, VA: National Council of Teachers of Mathematics.

40. Donoghue, Eileen F. 2003. The emergence of a profession: Mathematics education in the United States: 1890-1920. In *A history of school mathematics*, edited by G. M. A. Stanic and J. Kilpatrick. Reston, VA: National Council of Teachers of Mathematics.

41. Jones, Phillip S., and Arthur F. Coxford, eds. 1970. *A history of mathematics education in the United States and Canada*. Vol. 32nd Yearbook. Reston, VA: National Council of Teachers of Mathematics.

42. Stemhagen, Kurt. 2003. Toward a pragmatic/contextual philosophy of mathematics: Recovering Dewey's *Psychology of Number*. In *Philosophy of Education Yearbook*.

43. Good, Thomas L., ed. 2000. *American education: Yesterday, today, and tomorrow*. Chicago: The University of Chicago Press.

44. Jones, Phillip S., and Arthur F. Coxford, eds. 1970. *A history of mathematics education in the United States and Canada*. Vol. 32nd Yearbook. Reston, VA: National Council of Teachers of Mathematics.

45. Fey, James T. 1978. The United States experience with New Math. *Educational Studies in Mathematics* 9 (3):339-353.

46. Jones, Phillip S., and Arthur F. Coxford, eds. 1970. *A history of mathematics education in the United States and Canada*. Vol. 32nd Yearbook. Reston, VA: National Council of Teachers of Mathematics.

47. Stanic, George M.A., and Jeremy Kilpatrick, eds. 2003. *A history of school mathematics*. Vol. 1 & 2. Reston, VA: National Council of Teachers of Mathematics.

48. Current, Richard N., T. Harry Williams, Frank Freidel, and W. Elliot Brownlee. 1986. *The essentials of American history since 1865*. New York: Alfred A. Knopf, Inc.

49. State, United States Department of. 2006. *Outline of US history* [Internet] 2005 [cited January 18 2006]. Available from http://usinfo.state.gov/products/pubs/histryotln.

50. Ingui, Mary Jane Capozzoli. 1993. *American History 1877 to the Present*. Hauppauge, NY: Barron's

51. Current, Richard N., T. Harry Williams, Frank Freidel, and W. Elliot Brownlee. 1986. *The essentials of American history since 1865*. New York: Alfred A. Knopf, Inc.

52. Tozer, Steven E., Paul C. Violas, and Guy Senese. 1998. *School and society: Historical and contemporary perspectives*. Boston: McGraw Hill.

53. Current, Richard N., T. Harry Williams, Frank Freidel, and W. Elliot Brownlee. 1986. *The essentials of American history since 1865*. New York: Alfred A. Knopf, Inc.

54. Tozer, Steven E., Paul C. Violas, and Guy Senese. 1998. *School and society: Historical and contemporary perspectives*. Boston: McGraw Hill.

55. Rury, John L. 2005. *Education and social change: Themes in the history of American schooling*. 2nd ed. Mahwah, NJ: Lawrence Erlbaum Associates.

56. Klein, David. 2003. A brief history of American K-12 mathematics education in the 20th century. In *Mathematical Cognition*, edited by J. Royer. Greenwich, CT: Information Age Publishing.

57. Smith, Melinda Ann. 2004. Reconceptualizing mathematics education, College of Graduate Studies, Georgia Southern University.

58. Grouws, Douglas A., and Kristin J. Cebulla. 2000. Elementary and middle school mathematics at the crossroads. In *American education: Yesterday, today, and tomorrow*, edited by T. L. Good. Chicago: The University Press of Chicago.

59. Ravitch, Diane. 2000. *Left back: A century of failed school reforms*. New York: Simon & Schuster.

60. Smith, Melinda Ann. 2004. Reconceptualizing mathematics education, College of Graduate Studies, Georgia Southern University.

61. Grouws, Douglas A., and Kristin J. Cebulla. 2000. Elementary and middle school mathematics at the crossroads. In *American education: Yesterday, today, and tomorrow*, edited by T. L. Good. Chicago: The University Press of Chicago.

62. Smith, Melinda Ann. 2004. Reconceptualizing mathematics education, College of Graduate Studies, Georgia Southern University.

63. Kilpatrick, Jeremy. 1992. A history of research in mathematics education. In *Handbook of Research on Mathematics Teaching and Learning*, edited by D. A. Grouws. New York: Macmillan.

64. Jones, Eric D., and W. Thomas Southern. 2003. Balancing perspectives on mathematics instruction. *Focus on Exceptional Children* 35 (9):1-15.

65. Jones, Phillip S., and Arthur F. Coxford, eds. 1970. *A history of mathematics education in the United States and Canada*. Vol. 32nd Yearbook. Reston, VA: National Council of Teachers of Mathematics.

66. Ravitch, Diane. 2000. *Left back: A century of failed school reforms*. New York: Simon & Schuster.

67. Kilpatrick, Jeremy. 1992. A history of research in mathematics education. In *Handbook of Research on Mathematics Teaching and Learning*, edited by D. A. Grouws. New York: Macmillan.

68. Stanic, George M.A., and Jeremy Kilpatrick, eds. 2003. *A history of school mathematics*. Vol. 1 & 2. Reston, VA: National Council of Teachers of Mathematics.

69. Parker, Franklin. 1993. Turning points: Books and reports that reflected and shaped U.S. education, 1749-1990s, edited by ED369695: U.S. Department of Education.

70. Oliver, J. Steve, and B. Kim Nichols. 1999. Early Days. *School Science and Mathematics* 99 (2):102-105.

71. Parker, Franklin. 1993. Turning points: Books and reports that reflected and shaped U.S. education, 1749-1990s, edited by ED369695: U.S. Department of Education.

72.   State, United States Department of. 2006. *Outline of US history* [Internet] 2005 [cited January 18 2006]. Available from http://usinfo.state.gov/products/pubs/histryotln.

73.   Current, Richard N., T. Harry Williams, Frank Freidel, and W. Elliot Brownlee. 1986. *The essentials of American history since 1865*. New York: Alfred A. Knopf, Inc.

74.   Tozer, Steven E., Paul C. Violas, and Guy Senese. 1998. *School and society: Historical and contemporary perspectives*. Boston: McGraw Hill.

75.   Sadker, Myra P., and David M. Sadker. 2000. *Teachers, schools, and society*. Boston: McGraw Hill.

76.   Kliebard, Herbert M., and Barry M. Franklin. 2003. The ascendance of practical and vocational mathematics, 1893-1945: Academic mathematics under siege. In *A history of school mathematics*, edited by G. M. A. Stanic and J. Kilpatrick. Reston, VA: National Council of Teachers of Mathematics.

77.   Ravitch, Diane. 2000. *Left back: A century of failed school reforms*. New York: Simon & Schuster.

78.   Garrett, Alan W., and Jr. O.L. Davis. 2003. A time of uncertainty and change: School mathematics from World War II until the new math. In *A history of school mathematics*, edited by G. M. A. Stanic and J. Kilpatrick. Reston, VA: National Council of Teachers of Mathematics.

79.   Ravitch, Diane. 2000. *Left back: A century of failed school reforms*. New York: Simon & Schuster.

80.   Kilpatrick, Jeremy. 1992. A history of research in mathematics education. In *Handbook of Research on Mathematics Teaching and Learning*, edited by D. A. Grouws. New York: Macmillan.

81.   Grouws, Douglas A., and Kristin J. Cebulla. 2000. Elementary and middle school mathematics at the crossroads. In *American education: Yesterday, today, and tomorrow*, edited by T. L. Good. Chicago: The University Press of Chicago.

82.   NCTM. 1933. *The teaching of mathematics in the secondary school, The Eighth Yearbook*. New York: National Council of Teachers of Mathematics.

83.   Grinberg, Jaime Gerardo Alberto. 2002. "I had never been exposed to teaching like that": Progressive teacher education at Bank Street during the 1930s. *Teachers College Record* 104 (7):1422-1460.

84.   NCTM. 1933. *The teaching of mathematics in the secondary school, The Eighth Yearbook*. New York: National Council of Teachers of Mathematics.

85.   Jones, Phillip S., and Arthur F. Coxford, eds. 1970. *A history of mathematics education in the United States and Canada*. Vol. 32nd Yearbook. Reston, VA: National Council of Teachers of Mathematics.

86.   Ravitch, Diane. 2000. *Left back: A century of failed school reforms*. New York: Simon & Schuster.

87.   Jones, Phillip S., and Arthur F. Coxford, eds. 1970. *A history of mathematics education in the United States and Canada*. Vol. 32nd Yearbook. Reston, VA: National Council of Teachers of Mathematics.

88.   Shirley, Lawrence, and Iorhemen Kyeleve. 2005. A cyclic pattern of mathematics curriculum trends. In *Future Directions in Science, Mathematics, and Technical Education*. Universiti Brunei Darussalam.

89.   Jones, Phillip S., and Arthur F. Coxford, eds. 1970. *A history of mathematics education in the United States and Canada*. Vol. 32nd Yearbook. Reston, VA: National Council of Teachers of Mathematics.

90.   Sadker, Myra P., and David M. Sadker. 2000. *Teachers, schools, and society*. Boston: McGraw Hill.

91. Herrera, Terese A., and Douglas T. Owens. 2001. The "new new math"?: Two reform movements in mathematics education. *Theory into Practice* 40 (2):84-92.

92. Jones, Phillip S., and Arthur F. Coxford, eds. 1970. *A history of mathematics education in the United States and Canada*. Vol. 32nd Yearbook. Reston, VA: National Council of Teachers of Mathematics.

93. Tindall, George Brown, and David Emory Shi. 1999. *America: A narrative history*. New York: W.W. Norton.

94. Tozer, Steven E., Paul C. Violas, and Guy Senese. 1998. *School and society: Historical and contemporary perspectives*. Boston: McGraw Hill.

95. Dotzler, James J. 2003. A note on the nature and history of post-secondary developmental education in America. Garden City, NY: Nassau Community College (SUNY).

96. Garraty, John A., and Mark C. Carnes. 2000. *The American nation: A history of the United States since 1865*. New York: Addison Wesley Longman, Inc.

97. Tozer, Steven E., Paul C. Violas, and Guy Senese. 1998. *School and society: Historical and contemporary perspectives*. Boston: McGraw Hill.

98. Ravitch, Diane. 2000. *Left back: A century of failed school reforms*. New York: Simon & Schuster.

99. Raimi, Ralph A. 2000. Judging state standards for K-12 mathematics education. In *What's at stake in the K-12 standards wars*, edited by S. Stotsky. New York: Peter Lang.

100. NCTM. 1940. *The place of mathematics in secondary education, Fifteenth Yearbook*. New York: The Mathematical Association of America and the National Council of Teachers of Mathematics.

101. Garrett, Alan W., and Jr. O.L. Davis. 2003. A time of uncertainty and change: School mathematics from World War II until the new math. In *A history of school mathematics*, edited by G. M. A. Stanic and J. Kilpatrick. Reston, VA: National Council of Teachers of Mathematics.

102. Grouws, Douglas A., and Kristin J. Cebulla. 2000. Elementary and middle school mathematics at the crossroads. In *American education: Yesterday, today, and tomorrow*, edited by T. L. Good. Chicago: The University Press of Chicago.

103. Kliebard, Herbert M., and Barry M. Franklin. 2003. The ascendance of practical and vocational mathematics, 1893-1945: Academic mathematics under siege. In *A history of school mathematics*, edited by G. M. A. Stanic and J. Kilpatrick. Reston, VA: National Council of Teachers of Mathematics.

104. Garrett, Alan W., and Jr. O.L. Davis. 2003. A time of uncertainty and change: School mathematics from World War II until the new math. In *A history of school mathematics*, edited by G. M. A. Stanic and J. Kilpatrick. Reston, VA: National Council of Teachers of Mathematics.

105. Tozer, Steven E., Paul C. Violas, and Guy Senese. 1998. *School and society: Historical and contemporary perspectives*. Boston: McGraw Hill.

106. Jones, Phillip S., and Arthur F. Coxford, eds. 1970. *A history of mathematics education in the United States and Canada*. Vol. 32nd Yearbook. Reston, VA: National Council of Teachers of Mathematics.

107. NCTM. 1940. *The place of mathematics in secondary education, Fifteenth Yearbook*. New York: The Mathematical Association of America and the National Council of Teachers of Mathematics.

108. Stanic, George M.A., and Jeremy Kilpatrick, eds. 2003. *A history of school mathematics*. Vol. 1 & 2. Reston, VA: National Council of Teachers of Mathematics.

109.   Ravitch, Diane. 2000. *Left back: A century of failed school reforms*. New York: Simon & Schuster.

110.   Kilpatrick, Jeremy. 1992. A history of research in mathematics education. In *Handbook of Research on Mathematics Teaching and Learning*, edited by D. A. Grouws. New York: Macmillan.

111.   Smith, Melinda Ann. 2004. Reconceptualizing mathematics education, College of Graduate Studies, Georgia Southern University.

112.   Stanic, George M.A., and Jeremy Kilpatrick, eds. 2003. *A history of school mathematics*. Vol. 1 & 2. Reston, VA: National Council of Teachers of Mathematics.

113.   Jones, Phillip S., and Arthur F. Coxford, eds. 1970. *A history of mathematics education in the United States and Canada*. Vol. 32nd Yearbook. Reston, VA: National Council of Teachers of Mathematics.

114.   Raimi, Ralph A. 2000. Judging state standards for K-12 mathematics education. In *What's at stake in the K-12 standards wars*, edited by S. Stotsky. New York: Peter Lang.

115.   Loveless, Tom, ed. 2001. *The great curriculum debate*. Washington, DC: Brookings Institution Press.

116.   Fey, James T., and Anna O. Graeber. 2003. From the new math to the *Agenda for Action*. In *A history of school mathematics*, edited by G. M. A. Stanic and J. Kilpatrick. Reston, VA: National Council of Teachers of Mathematics.

117.   State, United States Department of. 2006. *Outline of US history* [Internet] 2005 [cited January 18 2006]. Available from http://usinfo.state.gov/products/pubs/histryotln.

118.   Tindall, George Brown, and David Emory Shi. 1999. *America: A narrative history*. New York: W.W. Norton.

119.   Rury, John L. 2005. *Education and social change: Themes in the history of American schooling*. 2nd ed. Mahwah, NJ: Lawrence Erlbaum Associates.

120.   Tindall, George Brown, and David Emory Shi. 1999. *America: A narrative history*. New York: W.W. Norton.

121.   Berube, Maurice R. 1994. *American school reform: Progressive, equity, and excellence movements, 1883-1993*. Westport, CT: Greenwood Press.

122.   Rury, John L. 2005. *Education and social change: Themes in the history of American schooling*. 2nd ed. Mahwah, NJ: Lawrence Erlbaum Associates.

123.   Tindall, George Brown, and David Emory Shi. 1999. *America: A narrative history*. New York: W.W. Norton.

124.   Garraty, John A., and Mark C. Carnes. 2000. *The American nation: A history of the United States since 1865*. New York: Addison Wesley Longman, Inc.

125.   Tozer, Steven E., Paul C. Violas, and Guy Senese. 1998. *School and society: Historical and contemporary perspectives*. Boston: McGraw Hill.

126.   Garraty, John A., and Mark C. Carnes. 2000. *The American nation: A history of the United States since 1865*. New York: Addison Wesley Longman, Inc.

127.   Ravitch, Diane. 2000. *Left back: A century of failed school reforms*. New York: Simon & Schuster.

128.   Rury, John L. 2005. *Education and social change: Themes in the history of American schooling*. 2nd ed. Mahwah, NJ: Lawrence Erlbaum Associates.

129.   Burrill, Gail. 2001. Mathematics education: the future and the past create a context for today's issues. In *The great curriculum debate*, edited by T. Loveless. Washington, DC: Brookings Institution Press.

130. Kilpatrick, Jeremy. 1992. A history of research in mathematics education. In *Handbook of Research on Mathematics Teaching and Learning*, edited by D. A. Grouws. New York: Macmillan.

131. Klein, David. 2003. A brief history of American K-12 mathematics education in the 20th century. In *Mathematical Cognition*, edited by J. Royer. Greenwich, CT: Information Age Publishing.

132. Ravitch, Diane. 2000. *Left back: A century of failed school reforms*. New York: Simon & Schuster.

133. Grouws, Douglas A., and Kristin J. Cebulla. 2000. Elementary and middle school mathematics at the crossroads. In *American education: Yesterday, today, and tomorrow*, edited by T. L. Good. Chicago: The University Press of Chicago.

134. NCTM. 1957. *Insights into modern mathematics, Twenty-third yearbook*. Washington, DC: National Council of Teachers of Mathematics.

135. Loveless, Tom, ed. 2001. *The great curriculum debate*. Washington, DC: Brookings Institution Press.

136. NACOME, National Advisory Committee on Mathematical Education. 1975. Overview and analysis of school mathematics, grades K-12. Reston, VA.

137. Stanic, George M.A., and Jeremy Kilpatrick, eds. 2003. *A history of school mathematics*. Vol. 1 & 2. Reston, VA: National Council of Teachers of Mathematics.

138. NCTM. 1957. *Insights into modern mathematics, Twenty-third yearbook*. Washington, DC: National Council of Teachers of Mathematics.

139. Good, Thomas L., ed. 2000. *American education: Yesterday, today, and tomorrow*. Chicago: The University of Chicago Press.

140. Stanic, George M.A., and Jeremy Kilpatrick, eds. 2003. *A history of school mathematics*. Vol. 1 & 2. Reston, VA: National Council of Teachers of Mathematics.

141. Smith, Melinda Ann. 2004. Reconceptualizing mathematics education, College of Graduate Studies, Georgia Southern University.

142. Sobel, Max, and Evan Maletsky. 1999. *Teaching mathematics: A sourcebook of aids, activities, and strategies*. Boston: Allyn & Bacon.

143. State, United States Department of. 2006. *Outline of US history* [Internet] 2005 [cited January 18 2006]. Available from http://usinfo.state.gov/products/pubs/histryotln.

144. Tindall, George Brown, and David Emory Shi. 1999. *America: A narrative history*. New York: W.W. Norton.

145. Angus, David L., and Jeffrey E. Mirel. 2003. Mathematics enrollments and the development of the high school in the United States, 1910-1994. In *A history of school mathematics*, edited by G. M. A. Stanic and J. Kilpatrick. Reston, VA: National Council of Teachers of Mathematics.

146. Morrison, George S. 2000. *Teaching in America*. Boston: Allyn & Bacon.

147. Ravitch, Diane. 2000. *Left back: A century of failed school reforms*. New York: Simon & Schuster.

148. Council, National Research. 1989. *Everybody counts: A report to the nation on the future of mathematics education*. Washington, DC: National Academy Press.

149. NACOME, National Advisory Committee on Mathematical Education. 1975. Overview and analysis of school mathematics, grades K-12. Reston, VA.

150. Fey, James T., and Anna O. Graeber. 2003. From the new math to the *Agenda for Action*. In *A history of school mathematics*, edited by G. M. A. Stanic and J. Kilpatrick. Reston, VA: National Council of Teachers of Mathematics.

151.   Stanic, George M.A., and Jeremy Kilpatrick, eds. 2003. *A history of school mathematics*. Vol. 1 & 2. Reston, VA: National Council of Teachers of Mathematics.

152.   Jones, Phillip S., and Arthur F. Coxford, eds. 1970. *A history of mathematics education in the United States and Canada*. Vol. 32nd Yearbook. Reston, VA: National Council of Teachers of Mathematics.

153.   Tozer, Steven E., Paul C. Violas, and Guy Senese. 1998. *School and society: Historical and contemporary perspectives*. Boston: McGraw Hill.

154.   Jones, Phillip S., and Arthur F. Coxford, eds. 1970. *A history of mathematics education in the United States and Canada*. Vol. 32nd Yearbook. Reston, VA: National Council of Teachers of Mathematics.

155.   Parker, Franklin. 1993. Turning points: Books and reports that reflected and shaped U.S. education, 1749-1990s, edited by ED369695: U.S. Department of Education.

156.   Walmsley, Angela L.E. 2003. *A history of the "new mathematics" movement and its relationship with current mathematical reform*. Lanham, MD: University Press of America.

157.   Raimi, Ralph A. 2000. Judging state standards for K-12 mathematics education. In *What's at stake in the K-12 standards wars*, edited by S. Stotsky. New York: Peter Lang.

158.   Jones, Phillip S., and Arthur F. Coxford, eds. 1970. *A history of mathematics education in the United States and Canada*. Vol. 32nd Yearbook. Reston, VA: National Council of Teachers of Mathematics.

159.   Smith, Melinda Ann. 2004. Reconceptualizing mathematics education, College of Graduate Studies, Georgia Southern University.

160.   Foster, John. 1972. *Discovery learning in the primary school*. London: Routledge & Kegan Paul.

161.   Fey, James T., and Anna O. Graeber. 2003. From the new math to the *Agenda for Action*. In *A history of school mathematics*, edited by G. M. A. Stanic and J. Kilpatrick. Reston, VA: National Council of Teachers of Mathematics.

162.   Stanic, George M.A., and Jeremy Kilpatrick, eds. 2003. *A history of school mathematics*. Vol. 1 & 2. Reston, VA: National Council of Teachers of Mathematics.

163.   Parker, Franklin. 1993. Turning points: Books and reports that reflected and shaped U.S. education, 1749-1990s, edited by ED369695: U.S. Department of Education.

164.   Jones, Phillip S., and Arthur F. Coxford, eds. 1970. *A history of mathematics education in the United States and Canada*. Vol. 32nd Yearbook. Reston, VA: National Council of Teachers of Mathematics.

165.   Ravitch, Diane. 2000. *Left back: A century of failed school reforms*. New York: Simon & Schuster.

166.   Parker, Franklin. 1993. Turning points: Books and reports that reflected and shaped U.S. education, 1749-1990s, edited by ED369695: U.S. Department of Education.

167.   Klein, David. 2003. A brief history of American K-12 mathematics education in the 20th century. In *Mathematical Cognition*, edited by J. Royer. Greenwich, CT: Information Age Publishing.

168.   Good, Thomas L., ed. 2000. *American education: Yesterday, today, and tomorrow*. Chicago: The University of Chicago Press.

169.   Fey, James T., and Anna O. Graeber. 2003. From the new math to the *Agenda for Action*. In *A history of school mathematics*, edited by G. M. A. Stanic and J. Kilpatrick. Reston, VA: National Council of Teachers of Mathematics.

170.   Herrera, Terese A., and Douglas T. Owens. 2001. The "new new math"?: Two reform movements in mathematics education. *Theory into Practice* 40 (2):84-92.

171.   Fey, James T.   1978.   The United States experience with New Math. *Educational Studies in Mathematics* 9 (3):339-353.

172.   Fey, James T., and Anna O. Graeber. 2003. From the new math to the *Agenda for Action*. In *A history of school mathematics*, edited by G. M. A. Stanic and J. Kilpatrick. Reston, VA: National Council of Teachers of Mathematics.

173.   Stanic, George M.A., and Jeremy Kilpatrick, eds. 2003. *A history of school mathematics*. Vol. 1 & 2. Reston, VA: National Council of Teachers of Mathematics.

174.   Snider, Robert.   1978.   Back to the basics? Washington, DC: National Education Association.

175.   Wu, H.   1996.   The mathematician and the mathematics education reform. *Notices of the AMS* 43 (12):1531-1537.

176.   Tindall, George Brown, and David Emory Shi. 1999. *America: A narrative history*. New York: W.W. Norton.

177.   State, United States Department of. 2006. *Outline of US history* [Internet] 2005 [cited January 18 2006]. Available from http://usinfo.state.gov/products/pubs/histryotln.

178.   Tindall, George Brown, and David Emory Shi. 1999. *America: A narrative history*. New York: W.W. Norton.

179.   Ingui, Mary Jane Capozzoli. 1993. *American History 1877 to the Present*. Hauppauge, NY: Barron's

180.   Tindall, George Brown, and David Emory Shi. 1999. *America: A narrative history*. New York: W.W. Norton.

181.   Rury, John L. 2005. *Education and social change: Themes in the history of American schooling*. 2nd ed. Mahwah, NJ: Lawrence Erlbaum Associates.

182.   Ravitch, Diane. 2000. *Left back: A century of failed school reforms*. New York: Simon & Schuster.

183.   Snider, Robert.   1978.   Back to the basics? Washington, DC: National Education Association.

184.   Angus, David L., and Jeffrey E. Mirel. 2003. Mathematics enrollments and the development of the high school in the United States, 1910-1994. In *A history of school mathematics*, edited by G. M. A. Stanic and J. Kilpatrick. Reston, VA: National Council of Teachers of Mathematics.

185.   Grouws, Douglas A., and Kristin J. Cebulla. 2000. Elementary and middle school mathematics at the crossroads. In *American education: Yesterday, today, and tomorrow*, edited by T. L. Good. Chicago: The University Press of Chicago.

186.   Fey, James T., and Anna O. Graeber. 2003. From the new math to the *Agenda for Action*. In *A history of school mathematics*, edited by G. M. A. Stanic and J. Kilpatrick. Reston, VA: National Council of Teachers of Mathematics.

187.   Grouws, Douglas A., and Kristin J. Cebulla. 2000. Elementary and middle school mathematics at the crossroads. In *American education: Yesterday, today, and tomorrow*, edited by T. L. Good. Chicago: The University Press of Chicago.

188.   Burrill, Gail. 2001. Mathematics education: the future and the past create a context for today's issues. In *The great curriculum debate*, edited by T. Loveless. Washington, DC: Brookings Institution Press.

189.   Grouws, Douglas A., and Kristin J. Cebulla. 2000. Elementary and middle school mathematics at the crossroads. In *American education: Yesterday,*

today, and tomorrow, edited by T. L. Good. Chicago: The University Press of Chicago.

190.   Tozer, Steven E., Paul C. Violas, and Guy Senese. 1998. *School and society: Historical and contemporary perspectives*. Boston: McGraw Hill.

191.   Stanic, George M.A., and Jeremy Kilpatrick, eds. 2003. *A history of school mathematics*. Vol. 1 & 2. Reston, VA: National Council of Teachers of Mathematics.

192.   NACOME, National Advisory Committee on Mathematical Education. 1975. Overview and analysis of school mathematics, grades K-12. Reston, VA.

193.   Ball, Deborah Loewenberg, and Hyman Bass. 2000. Interweaving content and pedagogy in teaching and learning to teach: Knowing and using mathematics. In *Knowledge and power in the global economy: Politics and the rhetoric of school reform*, edited by D. A. Gabbard. Mahwah, NJ: Lawrence Erlbaum Associates.

194.   Ravitch, Diane. 2000. *Left back: A century of failed school reforms*. New York: Simon & Schuster.

195.   Fey, James T., and Anna O. Graeber. 2003. From the new math to the *Agenda for Action*. In *A history of school mathematics*, edited by G. M. A. Stanic and J. Kilpatrick. Reston, VA: National Council of Teachers of Mathematics.

196.   Taylor, P. Mark. 2002. Implementing the standards: Keys to establishing positive professional inertia in preservice mathematics teachers. *School Science and Mathematics* 102 (3):137-141.

197.   Herrera, Terese A., and Douglas T. Owens. 2001. The "new new math"?: Two reform movements in mathematics education. *Theory into Practice* 40 (2):84-92.

198.   Ravitch, Diane. 2000. *Left back: A century of failed school reforms*. New York: Simon & Schuster.

199.   Good, Thomas L., ed. 2000. *American education: Yesterday, today, and tomorrow*. Chicago: The University of Chicago Press.

200.   Stanic, George M.A., and Jeremy Kilpatrick, eds. 2003. *A history of school mathematics*. Vol. 1 & 2. Reston, VA: National Council of Teachers of Mathematics.

201.   Grouws, Douglas A., and Kristin J. Cebulla. 2000. Elementary and middle school mathematics at the crossroads. In *American education: Yesterday, today, and tomorrow*, edited by T. L. Good. Chicago: The University Press of Chicago.

202.   Stanic, George M.A., and Jeremy Kilpatrick, eds. 2003. *A history of school mathematics*. Vol. 1 & 2. Reston, VA: National Council of Teachers of Mathematics.

203.   Tindall, George Brown, and David Emory Shi. 1999. *America: A narrative history*. New York: W.W. Norton.

204.   State, United States Department of. 2006. *Outline of US history* [Internet] 2005 [cited January 18 2006]. Available from http://usinfo.state.gov/products/pubs/histryotln.

205.   Tindall, George Brown, and David Emory Shi. 1999. *America: A narrative history*. New York: W.W. Norton.

206.   Stanic, George M.A., and Jeremy Kilpatrick, eds. 2003. *A history of school mathematics*. Vol. 1 & 2. Reston, VA: National Council of Teachers of Mathematics.

207.   Government, US. 1983. *A nation at risk: The imperative for educational reform*. Washington, DC: The Commission on Excellence in Education.

208. Ravitch, Diane. 2000. *Left back: A century of failed school reforms.* New York: Simon & Schuster.

209. Grouws, Douglas A., and Kristin J. Cebulla. 2000. Elementary and middle school mathematics at the crossroads. In *American education: Yesterday, today, and tomorrow*, edited by T. L. Good. Chicago: The University Press of Chicago.

210. NCTM. 1980. *An agenda for action.* Reston, VA: National Council of Teachers of Mathematics.

211. Stanic, George M.A., and Jeremy Kilpatrick, eds. 2003. *A history of school mathematics.* Vol. 1 & 2. Reston, VA: National Council of Teachers of Mathematics.

212. Grouws, Douglas A., and Kristin J. Cebulla. 2000. Elementary and middle school mathematics at the crossroads. In *American education: Yesterday, today, and tomorrow*, edited by T. L. Good. Chicago: The University Press of Chicago.

213. NCTM. 1980. *An agenda for action.* Reston, VA: National Council of Teachers of Mathematics.

214. Shirley, Lawrence, and Iorhemen Kyeleve. 2005. A cyclic pattern of mathematics curriculum trends. In *Future Directions in Science, Mathematics, and Technical Education.* Universiti Brunei Darussalam.

215. Ravitch, Diane. 2000. *Left back: A century of failed school reforms.* New York: Simon & Schuster.

216. Tozer, Steven E., Paul C. Violas, and Guy Senese. 1998. *School and society: Historical and contemporary perspectives.* Boston: McGraw Hill.

217. Stanic, George M.A., and Jeremy Kilpatrick, eds. 2003. *A history of school mathematics.* Vol. 1 & 2. Reston, VA: National Council of Teachers of Mathematics.

218. NCTM. 1989. *Curriculum standards for school mathematics.* Reston, VA: National Council of Teachers of Mathematics.

219. Grouws, Douglas A., and Kristin J. Cebulla. 2000. Elementary and middle school mathematics at the crossroads. In *American education: Yesterday, today, and tomorrow*, edited by T. L. Good. Chicago: The University Press of Chicago.

220. Council, National Research. 1989. *Everybody counts: A report to the nation on the future of mathematics education.* Washington, DC: National Academy Press.

221. Stanic, George M.A., and Jeremy Kilpatrick, eds. 2003. *A history of school mathematics.* Vol. 1 & 2. Reston, VA: National Council of Teachers of Mathematics.

222. Grouws, Douglas A., and Kristin J. Cebulla. 2000. Elementary and middle school mathematics at the crossroads. In *American education: Yesterday, today, and tomorrow*, edited by T. L. Good. Chicago: The University Press of Chicago.

223. Marshall, John. 2003. Math wars: Taking sides. *Phi Delta Kappan*:193-249.

224. Parker, Franklin. 1993. Turning points: Books and reports that reflected and shaped U.S. education, 1749-1990s, edited by ED369695: U.S. Department of Education.

225. Curcio, Frances R. 1999. Dispelling myths about reform in school mathematics. *Mathematics Teaching in the Middle School* 4 (5):282-284.

226. Sobel, Max, and Evan Maletsky. 1999. *Teaching mathematics: A sourcebook of aids, activities, and strategies.* Boston: Allyn & Bacon.

227. Tindall, George Brown, and David Emory Shi. 1999. *America: A narrative history*. New York: W.W. Norton.

228. Garraty, John A., and Mark C. Carnes. 2000. *The American nation: A history of the United States since 1865*. New York: Addison Wesley Longman, Inc.

229. Tozer, Steven E., Paul C. Violas, and Guy Senese. 1998. *School and society: Historical and contemporary perspectives*. Boston: McGraw Hill.

230. Klein, David. 2003. A brief history of American K-12 mathematics education in the 20th century. In *Mathematical Cognition*, edited by J. Royer. Greenwich, CT: Information Age Publishing.

231. Grouws, Douglas A., and Kristin J. Cebulla. 2000. Elementary and middle school mathematics at the crossroads. In *American education: Yesterday, today, and tomorrow*, edited by T. L. Good. Chicago: The University Press of Chicago.

232. Klein, David. 2003. A brief history of American K-12 mathematics education in the 20th century. In *Mathematical Cognition*, edited by J. Royer. Greenwich, CT: Information Age Publishing.

233. Grouws, Douglas A., and Kristin J. Cebulla. 2000. Elementary and middle school mathematics at the crossroads. In *American education: Yesterday, today, and tomorrow*, edited by T. L. Good. Chicago: The University Press of Chicago.

234. Burrill, Gail. 2001. Mathematics education: the future and the past create a context for today's issues. In *The great curriculum debate*, edited by T. Loveless. Washington, DC: Brookings Institution Press.

235. Battista, Michael T. 2001. Research and reform in mathematics education. In *The great curriculum debate*, edited by T. Loveless. Washington, DC: Brookings Institution Press.

236. Good, Thomas L., ed. 2000. *American education: Yesterday, today, and tomorrow*. Chicago: The University of Chicago Press.

237. Grouws, Douglas A., and Kristin J. Cebulla. 2000. Elementary and middle school mathematics at the crossroads. In *American education: Yesterday, today, and tomorrow*, edited by T. L. Good. Chicago: The University Press of Chicago.

238. Hardy, Lawrence. 2005. The future of education: Not all we hoped, or had hyped. *Education Digest* 4-9.

239. Grouws, Douglas A., and Kristin J. Cebulla. 2000. Elementary and middle school mathematics at the crossroads. In *American education: Yesterday, today, and tomorrow*, edited by T. L. Good. Chicago: The University Press of Chicago.

240. Tozer, Steven E., Paul C. Violas, and Guy Senese. 1998. *School and society: Historical and contemporary perspectives*. Boston: McGraw Hill.

241. Ravitch, Diane. 2000. *Left back: A century of failed school reforms*. New York: Simon & Schuster.

242. Klein, David. 2003. A brief history of American K-12 mathematics education in the 20th century. In *Mathematical Cognition*, edited by J. Royer. Greenwich, CT: Information Age Publishing.

243. Ravitch, Diane. 2000. *Left back: A century of failed school reforms*. New York: Simon & Schuster.

244. Raimi, Ralph A. 2000. Judging state standards for K-12 mathematics education. In *What's at stake in the K-12 standards wars*, edited by S. Stotsky. New York: Peter Lang.

245.   Draper, Roni Jo. 2002. School mathematics reform, constructivism, and literacy: A case for literacy instruction in the reform-oriented math classroom. *Journal of Adolescent & Adult Literacy* 45 (6):520-523.

246.   Herrera, Terese A., and Douglas T. Owens. 2001. The "new new math"?: Two reform movements in mathematics education. *Theory into Practice* 40 (2):84-92.

247.   Marshall, John. 2003. Math wars: Taking sides. *Phi Delta Kappan*:193-249.

248.   Morgan, Candia. 2000. Better assessment in mathematics education? A social perspective. In *Knowledge and power in the global economy: Politics and the rhetoric of school reform*, edited by D. A. Gabbard. Mahway, NJ: Lawrence Erlbaum Associates.

249.   Herrera, Terese A., and Douglas T. Owens. 2001. The "new new math"?: Two reform movements in mathematics education. *Theory into Practice* 40 (2):84-92.

250.   Burrill, Gail. 2001. Mathematics education: the future and the past create a context for today's issues. In *The great curriculum debate*, edited by T. Loveless. Washington, DC: Brookings Institution Press.

251.   Schoen, Harold L., James T. Fey, Christian R. Hirsch, and Arhtur F. Coxford. 1999. Issues and options in the Math Wars. *Phi Delta Kappan* 80 (6):444-453.

252.   Angus, David L., and Jeffrey E. Mirel. 2003. Mathematics enrollments and the development of the high school in the United States, 1910-1994. In *A history of school mathematics*, edited by G. M. A. Stanic and J. Kilpatrick. Reston, VA: National Council of Teachers of Mathematics.

253.   Marshall, John. 2003. Math wars: Taking sides. *Phi Delta Kappan*:193-249.

254.   Tozer, Steven E., Paul C. Violas, and Guy Senese. 1998. *School and society: Historical and contemporary perspectives*. Boston: McGraw Hill.

255.   Sadker, Myra P., and David M. Sadker. 2000. *Teachers, schools, and society*. Boston: McGraw Hill.

256.   Loveless, Tom, ed. 2001. *The great curriculum debate*. Washington, DC: Brookings Institution Press.

257.   Klein, David. 2003. A brief history of American K-12 mathematics education in the 20th century. In *Mathematical Cognition*, edited by J. Royer. Greenwich, CT: Information Age Publishing.

258.   Garelick, Barry. 2005. The new, A-maze-ing approach to math. *Education Next* 5 (2):28-36.

259.   Fey, James T., and Anna O. Graeber. 2003. From the new math to the *Agenda for Action*. In *A history of school mathematics*, edited by G. M. A. Stanic and J. Kilpatrick. Reston, VA: National Council of Teachers of Mathematics.

260.   Loveless, Tom, ed. 2001. *The great curriculum debate*. Washington, DC: Brookings Institution Press.

261.   Fey, James T., and Anna O. Graeber. 2003. From the new math to the *Agenda for Action*. In *A history of school mathematics*, edited by G. M. A. Stanic and J. Kilpatrick. Reston, VA: National Council of Teachers of Mathematics.

262.   Schoen, Harold L., James T. Fey, Christian R. Hirsch, and Arhtur F. Coxford. 1999. Issues and options in the Math Wars. *Phi Delta Kappan* 80 (6):444-453.

263.   Stanic, George M.A., and Jeremy Kilpatrick, eds. 2003. *A history of school mathematics*. Vol. 1 & 2. Reston, VA: National Council of Teachers of Mathematics.

264.   Jones, Phillip S., and Arthur F. Coxford, eds. 1970. *A history of mathematics education in the United States and Canada*. Vol. 32nd Yearbook. Reston, VA: National Council of Teachers of Mathematics.

265. Stanic, George M.A., and Jeremy Kilpatrick, eds. 2003. *A history of school mathematics*. Vol. 1 & 2. Reston, VA: National Council of Teachers of Mathematics.

266. Curcio, Frances R. 1999. Dispelling myths about reform in school mathematics. *Mathematics Teaching in the Middle School* 4 (5):282-284.

267. Stanic, George M.A., and Jeremy Kilpatrick, eds. 2003. *A history of school mathematics*. Vol. 1 & 2. Reston, VA: National Council of Teachers of Mathematics.

268. Loveless, Tom, ed. 2001. *The great curriculum debate*. Washington, DC: Brookings Institution Press.

269. Stein, S.K. 1996. *Strength in numbers: Discovering the joy and power of mathematics in everyday life*. New York: John Wiley & Sons, Inc.

270. Klein, David. 2003. A brief history of American K-12 mathematics education in the 20th century. In *Mathematical Cognition*, edited by J. Royer. Greenwich, CT: Information Age Publishing.

271. Loveless, Tom, ed. 2001. *The great curriculum debate*. Washington, DC: Brookings Institution Press.

272. NACOME, National Advisory Committee on Mathematical Education. 1975. Overview and analysis of school mathematics, grades K-12. Reston, VA.

273. Wu, H. 1999. Basic skills versus conceptual understanding. *American Educator*:1-7.

274. Schoen, Harold L., James T. Fey, Christian R. Hirsch, and Arhtur F. Coxford. 1999. Issues and options in the Math Wars. *Phi Delta Kappan* 80 (6):444-453.

275. Jones, Phillip S., and Arthur F. Coxford, eds. 1970. *A history of mathematics education in the United States and Canada*. Vol. 32nd Yearbook. Reston, VA: National Council of Teachers of Mathematics.

276. See note 275.

277. Battista, Michael T. 2001. Research and reform in mathematics education. In *The great curriculum debate*, edited by T. Loveless. Washington, DC: Brookings Institution Press.

278. Jones, Eric D., and W. Thomas Southern. 2003. Balancing perspectives on mathematics instruction. *Focus on Exceptional Children* 35 (9):1-15.

279. Fey, James T., and Anna O. Graeber. 2003. From the new math to the *Agenda for Action*. In *A history of school mathematics*, edited by G. M. A. Stanic and J. Kilpatrick. Reston, VA: National Council of Teachers of Mathematics.

280. Parker, Franklin. 1993. Turning points: Books and reports that reflected and shaped U.S. education, 1749-1990s, edited by ED369695: U.S. Department of Education.

281. Herrera, Terese A., and Douglas T. Owens. 2001. The "new new math"?: Two reform movements in mathematics education. *Theory into Practice* 40 (2):84-92.

282. Jones, Phillip S., and Arthur F. Coxford, eds. 1970. *A history of mathematics education in the United States and Canada*. Vol. 32nd Yearbook. Reston, VA: National Council of Teachers of Mathematics.

283. Marshall, John. 2003. Math wars: Taking sides. *Phi Delta Kappan*:193-249.

Index

multiple-choice tests, 37

National Advisory Committee on
  Mathematical Education, 36
National Aeronautics and Space
  Administration, 24
National Assessment of Educational
  Progress, 38, 48
National Commission on Excellence in
  Education, 39
National Council of Supervisors of
  Mathematics, 40,42
National Council of Teachers of
  Mathematics, 9, 13, 20, 21, 37, 39,
  41, 45,
National Education Association, 20
National Science Board's Science and
  Engineering Indicators, 46
NCTM Standards, 50
New Deal, 15, 16
new math, 19, 23, 25-33, 35-38, 41, 43,
  45, 50, 55, 56,
new mathematics, 16, 23, 33,
Nixon, Richard , 35
No Child Left Behind Act, 44
normalcy, 11

Peace Corps, 30
pedagogical techniques, 49
pedagogy, 5, 9, 13, 17, 22, 26, 32, 37,
  42, 41, 46, 55
philosophy, 1-3, 8, 11, 15, 16, 20, 24,
  30, 36, 40, 44, 55
Piaget, 32
practical mathematics, 20, 21
*Principles and Standards for School
  Mathematics*, 43
professional development, 41,50
professionalism, 9, 17
progressive education, 4, 5, 8,9, 11-13,
  16, 17, 22, 24, 25, 32, 42, 47-49, 50,
  55
Progressive Education Association, 9,
  16, 17, 22
progressive educators, 15, 17
progressive era, 3, 8, 50, 55
progressive ideals, 8
progressive movement, 3, 7, 8, 9, 12,
  46, 49, 50
progressive pedagogy, 5, 22
progressive reform, 49, 50
progressivism, 5, 9, 15, 46, 50, 55

prohibition, 3, 11
project-based learning, 50
psychological disorders, 35
psychological testing, 10
public education, 2, 16, 17

reform, 1, 17, 25, 27, 30, 47, 54. 55
reform movements, 53, 54
repeating patterns, 2
Republicans, 11, 39
roaring twenties, 11
Reagan, Ronald, 39
Russians, 28

School Mathematics Study Group, 23,
  29
secondary mathematics, 12, 20-22
secondary schools, 8, 9, 12
secondary teachers, 9, 29
September 11, 2001, 44
set theory, 26
skilled workers, 36
social efficiency movement, 8, 12
social programs, 27
Social Reconstructionism, 15
Soviet Union, 29, 43
space race, 29
Sputnik, 23, 24, 25, 29, 32, 43, 50
standardized test scores, 37
standardized testing, 5, 37, 38, 41, 47
Standards, 43, 45, 46
Standards Based, 44
Strategic Defense Initiative, 39
student centered classroom, 46

teacher accountability, 37
teacher certification, 22, 29, 66
teacher competence, 37
teacher education, 2, 17, 30, 54
teacher education, 5, 8, 12, 17, 21, 25,
  29,31, 36, 37, 41, 45, 46, 54
teacher preparation, 41
teacher training, 45
teachers, 1, 5, 9, 12, 13, 17, 20-22, 26,
  30-33, 36, 37, 39-42, 44- 47, 50, 53-
  56,
textbook-driven curriculum, 53
Title I, 28, 33, 37
traditional education, 2
traditional reform, 56
training methods, 13

Author Biography

Angela L.E. Walmsley, Ph.D. was born in central Illinois. It is here that she first experienced the beauty of mathematics with two mathematics teachers that inspired her in junior high and high school. From there, she attended the University of Illinois at Urbana-Champaign where she received her B.S. in mathematics with teacher certification in mathematics and general science. During the course of her studies, she attended the University of Dundee in Dundee, Scotland. This experience brought her to continue her education at Trinity College: The University of Dublin where she received a Diploma in Statistics and a Master in Education degree specializing in mathematics education. After teaching middle and secondary mathematics in Belfast, Northern Ireland, she moved to St. Louis, Missouri where she studied mathematics education again, receiving her Ph.D. in curriculum and instruction from Saint Louis University. She has taught pre-service teachers and various undergraduates at two universities in the area, and currently is a professor at St. Louis University in the Department of Research Methodology acting as a liaison to the education departments in the college. Undergraduate courses she has taught include general education, mathematics methods for elementary, middle, and secondary schools, mathematics for elementary and middle school teachers, inferential statistics, and introductory research methods. Graduate courses she has taught include applied research, capstone projects, general research methods, and qualitative research. Dr. Walmsley's interests include mathematics education, educational research, and research in schools. She happily lives in St. Louis, Missouri with her husband and three children.